T0220196

Springer Praxis Books

Popular Astronomy

The Springer Praxis Popular Astronomy series welcomes anybody with a passion for the night sky. Requiring no formal background in the sciences and including very little to no mathematics, these enriching reads will appeal to general readers and seasoned astronomy enthusiasts alike.

Many of the books in this series are well illustrated, with lavish figures, photographs, and maps. They are written in a highly accessible and engaging style that readers of popular science magazines can easily grasp, breaking down denser aspects of astronomy and its related fields to a digestible level.

From ancient cosmology to the latest astronomical discoveries, these books will enlighten, educate, and expand your interests far beyond the telescope.

Massimo Capaccioli

Red Moon

The Soviet Conquest of Space

Springer

Published in association with
Praxis Publishing
Chichester, UK

Massimo Capaccioli
Department of Physics (Emeritus)
University of Naples Federico II
Naples, Italy

Springer Praxis Books
ISSN 2626-8760   ISSN 2626-8779   (electronic)
Popular Astronomy
ISBN 978-3-031-54759-1   ISBN 978-3-031-54760-7   (eBook)
https://doi.org/10.1007/978-3-031-54760-7

Translation from the Italian language edition: "Luna Rossa. La Conquista Sovietica dello Spazio" by Massimo Capaccioli, © Carocci Editore 2019. Published by Carocci Editore. All Rights Reserved.

This Springer imprint is published by the registered company Springer Nature Switzerland AG
The registered company address is: Gewerbestrasse 11, 6330 Cham, Switzerland

Paper in this product is recyclable.

What is life? An illusion, a shadow, a story.
And the greatest good is little enough;
for all life is a dream,
and dreams themselves are only dreams.
Pedro Calderón de la Barca

To the memory of Prof. Mikhail V. Sazhin,
a great physicist and an unforgettable friend

Contents

About the Author

Massimo Capaccioli is an internationally renowned astrophysicist. He has taught at the Universities of Padua and Naples, where he is now the professor emeritus, and has published more than 500 scientific articles on extragalactic and cosmological topics and a number of textbooks, including *Foundations of Celestial Mechanics* (Springer 2022, with E. Bannikova). A publicist and popularizer, he has also written several educational books including *Arminio Nobile and the Measurement of the Sky, or the Misadventures of a Neapolitan Astronomer* (Springer, 2012, with S. Galano, in Italian), *Red Moon. Soviet conquest of space* (Carocci 2019, in Italian; Technosphera 2021, in Russian), *The Enchantment of Urania: 25 Centuries of Observation of the Sky* (Carocci 2020, in Italian; World Scientific, 2024), *The Sun and the Other Stars of Dante Alighieri. A Cosmographic Journey through the Divina Commedia* (World Scientific, 2022, with S. di Serego Alighieri), and *Flashes of genius. New stories of men and stars* (Carocci 2023, in Italian).

Ballad of the Moon Moon

The boy looks and looks the boy looks at the Moon.
Federico Garcia Lorca
The Moon is the first milestone on the road to the stars.
Arthur C. Clarke.

In mid-December 2017, Donald Trump signed a directive (White House Space Policy Directive-1) to refocus NASA's space program on research and exploration entrusted to both humans and robots. A move—the American president declared—that "*marks a first step in returning American astronauts to the Moon for the first time since 1972, for long-term exploration and use. This time, we will not only plant our flag and leave our footprints—we will establish a foundation for an eventual mission to Mars, and perhaps someday, to many worlds beyond*". These beautiful words would leave hope for an altruistic use of significant economic resources and ingenuity "*to follow virtue and knowledge*".[1] A revival of the American dream that so misled us during the hot years of reconstruction and the cold season of the creeping war between the U.S. and the USSR.

Just six months after this promising statement, however, during a meeting of his National Space Council in the East Room of the White House to sign Space Policy Directive-3, Trump dropped the mask and addressed the issue of military occupation of space with his usual arrogant frankness.

---

[1] The reasons given by Odysseus, in canto 26 of Dante's *Inferno*, to convince his reluctant companions to venture into the uncharted sea beyond the Pillars of Hercules.

© The Author(s), under exclusive license to Springer Nature
Switzerland AG 2024
M. Capaccioli, *Red Moon*, Springer Praxis Books,
https://doi.org/10.1007/978-3-031-54760-7_1

The essence of the American character is to explore new horizons and to tame new frontiers. But our destiny, beyond the Earth, is not only a matter of national identity, but a matter of national security. So important for our military [...] When it comes to defending America, it is not enough to merely have an American presence in space. We must have American dominance in space.

So much so that a few months after this statement, he ordered the Department of Defense and the Pentagon to "*immediately begin the process necessary to establish a space force as the sixth branch of the armed forces*". The motivation is summed up in a famous Latin phrase by an unknown author: "*Si vis pacem para bellum*", if you want peace, prepare for war. It was this mother of all things,[2] both hot and Cold War, that took us to the Moon, and scientific curiosity will certainly not be the only drive to take us back there within this decade. It is better not to deceive ourselves, because the game is serious and it is advisable to play it, or at least to observe it, keeping enough lucidity to avoid unpleasant awakenings. An attention that is becoming urgent these days, with the resumption of lunar missions not only by America and Russia, but by many other nations, with even private parties entering the business.

The last Apollo mission took place in December 1972, and on the Soviet side, the *Luna* missions continued until August 1976. Then it became crystal clear that the game was not worth the candle for any of the competitors. The budget could have been better spent on colonizing space around Earth and exploring interplanetary space with robotic probes, perhaps by pooling resources instead of competing head-on. So the Moon was forgotten for decades. It was too far away for manned missions and too close for further unmanned exploration. But recently, after a long hiatus, there has been renewed interest, again from the United States and the Russian Federation (natural heir to the USSR's experience), and also from the People's Republic of China, Israel, India, South Korea, Italy, Japan, the Emirates, and collectively for Europe, the European Space Agency. No new human being has yet set foot again on our satellite, but it can be expected to happen soon, perhaps under the banner of a pool of nations and through the capital and labor of private companies. What will happen?

As always, to try to understand we could ask for the help of history, "*witness of the times, the light of truth, the life of memory, the teacher of life,*

---

[2] "*War is the mother of all and the king of all*" is one of the famous sayings of Heraclitus. Born in the city of Ephesus on the coast of Ionia in the sixth century BC, the ancient Greek pre-Socratic philosopher placed competition at the center of his thinking.

the messenger of antiquity.[3] But history, especially that which is closest to us, is manipulable and manipulated, littered like a minefield with fake news. It is useless, if not harmful, if we do not try to understand its thread, to cleanse it of clichés and lies, and to rescue from oblivion people and facts that have been unjustly and speciously forgotten. No one has the recipe to do what only time can do and often not completely.

That's why, as we celebrate the 55th anniversary of Apollo 11's historic journey to the Moon and the first human footprint on the dust of our satellite, it's worth remembering how much this extraordinary, truly American enterprise owes to the "*communist enemy*" and to those who, in the 1960s, put the USSR far ahead in the space race. A competition more like medieval "*Palio di Siena*" than to a game in the name of fair play, motivated by reasons far different from the curiosity of a Jules Verne, a Giovanni Virginio Schiaparelli or a Percival Lowell, to name just a few of the scientists or fantasy writers who dreamed, and made us dream, of the immense celestial prairies open above the clouds.

Space had both a strategic value, for espionage, defense, and offense, and a strong promotional charge, precious in a season of confrontation to the death between two political, economic, and social systems equally young and strongly antagonistic: American liberalism, dear to capital, and Soviet-style communism, watched with interest and hope by the poor masses of the world. Two superpowers, victorious in a deadly world war, each as vast as a continent and each ready to assume and maintain world leadership, both playing hardball but avoiding direct confrontation on the battlefield, held back by nuclear fear. Fortunately, neither contender felt the need to utter: "*let Samson die with all the Philistines*".

Thus the armed conflict was transferred to minor wars, fought with traditional weapons and outside the home territories of the two main antagonists, and in a sublime game for the conquest of the Moon, played with attacks and responses, dunks and blocks, and innumerable feints, like at the ping-pong table and the green one of poker. The kind of tournament that delegated the defense of honor to a handful of champions representing the entire nation (much more prosaically, this is what happens today with national soccer teams).

---

[3] "*Historia vero testis temporum, lux veritatis, vita memoriae, magistra vitae, nuntia vetustatis*" (history is truly the witness of time, the light of truth, the life of memory, the director of life, the herald of antiquity), from *On the Orator* by Marcus Tullius Cicero, 55 BC.

But, since "*ex nihilo nihil fit*"[4] (nothing comes from nothing), both the Soviets and the Americans would have struggled much more to achieve their space goals without the help of the German scientists who flourished in Peenemünde around one of the secret weapons projects with which Hitler hoped to win the challenge against the world, in the evil belief of a "*Deutschland über Alles*" (Germany above all). From the island in the Baltic Sea, V2 rockets took off to ravage England and Belgium in the last years of the war. Silent arrows created to sow death and destruction, beautiful as Nibelung deities in a diabolical swan song in which genius mixes with the worst fate for a man, the loss of his humanity.

Then the weapons fell silent, or nearly so, to make room for the most terrible tug-of-war of all time, the Cold War between the West and the Soviet Union. Frightened by the Red Army, which they feared would take revenge for all the crimes committed by the Nazi aggressors, the Führer's scientists spontaneously surrendered to the Americans, bringing as a dowry the projects and some samples of the V2. A choice of field led by Wernher von Braun that not everyone could share. Those who were still hiding in Germany at the time of the overwhelming advance of the Russians were captured and spared to serve the strategic plans of Stalin.

Some crumbs of the rich cake also ended up in the hands of the British and the French. This was the starting point for the conception, albeit with considerable delay, of a joint European space program. An initiative of the scientific community that, together with the creation of the European Organization for Nuclear Research (CERN) and later the European Southern Observatory (ESO), anticipated the realization of one of the most beautiful dreams of the twentieth century, a united Europe, which politics and selfishness have managed to turn into an ugly duckling for the time being.

The story we will tell is one of great machines and ingenious devices, of scientific knowledge and boundless ambition, of sublime courage and self-denial, but also of jealousies, low blows, mistakes, and failures. A story of people, pawns on a chessboard, complicated by the traps of propaganda and the Soviet penchant for secrecy and partial truth. A story that is also mythology, rich in anecdotes and twists, legends and false news, as befits a saga. A story that begins far away, with dreams of flight and the Moon, and has its true roots in the last decades of the twentieth century, when science fiction transformed into science. A story set in the turbulent and complex panorama of the first seventy years of the twentieth century. We will retrace it until that fateful July 21, 1969, when, at 2:56 a.m. Greenwich time, the

---

[4] Dictum first argued by Parmenides, an ancient Greek pre-Socratic philosopher from Elea in Magna Graecia (Southern Italy).

American astronaut Neil Armstrong crossed the finish line of the last lap by placing his left foot on the surface of the Moon.

Since then, the space race has continued, with even more spectacular achievements, and continues today thanks to unusual and promising synergies between nations. The Moon, which disappeared from the objectives of the major space agencies in the early 1970s, is now returning, as we said, with different motivations and objectives, much more concrete than the simple competition for primacy. What will it lead to? Hard to say, but it is not unlikely that the next great war will be fought from space stations and lunar bases. Let's hope not, otherwise the risk is that the next one, as Einstein also said, "*will be fought with sticks and stones*".

This book was originally written for the Italian publisher Carocci at the request of its editor-in-chief, Gianluca Mori, to celebrate the 50th anniversary of the conquest of the Moon, and it appeared in Italian in 2019. It was also translated into Russian for the publisher Technosphera in 2021, with a wealth of notes by the Russian politician and jurist Yuri Baturin, who knows this subject well, having been a cosmonaut himself.

I took the opportunity of this English translation to update the original text, removing some ambiguities and errors and enriching the narrative with additional information. I wish to thank once again Luisa Castellani for her valuable contribution in editing the form and content of the Italian manuscript. Finally, I like to express a grateful thought to all those who put up with me while my head was buried in this book and I had no time for anything or anyone.

Naples, December 3, 2023

The Dawn

Earth is the cradle of humanity, but one cannot live in a cradle forever.
Konstantin E. Tsiolkovsky
I'm always more interested in the present and the future than the past.
Bernard Baruch

On the morning of October 5, 1957, the inhabitants of the Earth awoke
to discover with great astonishment that they had a second Moon. Not like
the one that has always accompanied lovers and travelers, poets and farmers,
in a rhythmic going of phases that make it a witness and measure of time,
but an artificial Moon: a small aluminum ball slightly larger than a water-
melon, barely distinguishable in the darkness of the night and yet talkative,
with its faint and obsessive beeping directed at everyone, without territorial
boundaries, of race and gender. The signal had been made deliberately easy to
pick up even with a simple radio device, so that anyone who wanted to could
hear it. The garrulous Moon had climbed up there, to hundreds of kilometers
from the crust of the planet inhabited by men, on the tip of a gigantic missile
designed for war, launched the day before from a secret base in Kazakhstan,
away from prying eyes, at 7 h and 29 min p.m., Greenwich time.[1]

---

[1] To obtain the official time of Moscow or Baikonur, where the Soviet (now Russian Federation)
launch base was located, simply add 3 or 6 h, respectively, to Greenwich time (also known as UTC,
Coordinated Universal Time).

© The Author(s), under exclusive license to Springer Nature
Switzerland AG 2024
M. Capaccioli, *Red Moon*, Springer Praxis Books,
https://doi.org/10.1007/978-3-031-54760-7_2

The press release from the Soviet government agency, the famous Tass (Telegraph Agency of the Soviet Union), crafted with consummate skill, reads:

For several years, scientific research and experimental design work have been conducted in the Soviet Union on the creation of artificial satellites of the Earth. [...] As a result of very intensive work by scientific research institutes and design bureaus the first artificial satellite in the world has been created. On October 4, 1957 this first satellite was successfully launched in the USSR. [...] At the present time the satellite is describing elliptical trajectories around the Earth, and its flight can be observed in the rays of the rising and setting Sun with the aid of very simple optical instruments (binoculars, telescopes, etc.). [...] According to calculations which now are being supplemented by direct observations, the satellite will travel at altitudes up to 900 kilometers above the surface of the Earth; the time for a complete revolution of the satellite will be one hour and thirty-five minutes; [...] Scientific stations located at various points in the Soviet Union are tracking the satellite and determining the elements of its trajectory. Since the density of the rarified upper layers of the atmosphere is not accurately known, there are no data at present for the precise determination of the satellite's lifetime and of the point of its entry into the dense layers of the atmosphere. Calculations have shown that owing to the tremendous velocity of the satellite, at the end of its existence it will burn up on reaching the dense layers of the atmosphere at an altitude of several tens of kilometers. As early as the end of the nineteenth century the possibility of realizing cosmic flights by means of rockets was first scientifically substantiated in Russia by the works of the outstanding Russian scientist, K.E. Tsiolkovsky. The successful launching of the first man-made Earth satellite makes a most important contribution to the treasure-house of world science and culture. The scientific experiment accomplished at such a great height is of tremendous importance for learning the properties of cosmic space and for studying the Earth as a planet of our Solar System. During the International Geophysical Year, the Soviet Union proposes launching several more artificial Earth satellites. These subsequent satellites will be larger and heavier and they will be used to carry out programs of scientific research. Artificial Earth satellites will pave the way to interplanetary travel and, apparently our contemporaries will witness how the freed and conscientious labor of the people of the new socialist society makes the most daring dreams of mankind a reality.[2]

There was great surprise around the world, especially because the small, useless Moon called *Sputnik*, which means *traveling companion*, spoke Russian instead of American English, as everyone in the West had expected. In fact, more than two years earlier, President Dwight Eisenhower, through his

---

[2] Reported by F.J. Krieger, *Behind the Sputniks, Public Affairs Press*, Washington, DC, 1958, pp. 311–12.

press secretary, had publicly announced the intention of the United States to launch small satellites into orbit to mark the upcoming International Geophysical Year. Given the economic power of the United States and the successes of Yankee science and technology in every field of human activity, no one could doubt that the promise would be kept, to the greater glory of our species and specifically of the new leaders of the planet.

Among other things, the Americans could count on the cooperation of Germany's brightest scientists and engineers, who had successfully designed and built the deadly V2 flying bombs, the pride of the Führer's arsenal of wonder weapons. In short, across the ocean in the New World, everything was in place to do it fast and do it well. Instead, it was the Soviets who struck first. Their surprise move triggered another muscle-flexing contest between the two superpowers and their respective ideologies: a very unusual challenge, entrusted to completely unconventional weapons but fully integrated into the broader framework of the Cold War.

We can mark the beginning of the Cold War with the meeting of the Big Three in the gardens of Cecilienhof Palace in Potsdam, in the part of Germany occupied by the Red Army, where Harry Truman, Joseph Stalin, and Winston Churchill met between July 17 and August 2, 1945, to negotiate peace and share the spoils of the defeated Third Reich. The already growing rivalry between the two formerly allied blocs degenerated two years later when Truman, to counter creeping Soviet expansionism, theorized the strategy of political and military aid to anti-communist forces everywhere. From then on, the Soviets and Americans would confront each other without direct clashes by fomenting real wars or supporting them economically and militarily in the Middle East, Africa, South America, Korea, China, Cuba, and Vietnam. It was as if the Second World War, instead of ending, had updated its actors, modes, and theaters.

The Russians won all the battles but lost the war. In fact, they triumphed for about ten years, humiliating the opponents, who seemed unable to keep up with the feared and cryptic Bolsheviks in the space race. But it was a Pyrrhic victory. Like the treacherous attack by Japanese fighter bombers on the naval base at Pearl Harbor, the spectacular Soviet successes provoked the reaction of a nation proud of its own role in the world and of an unparalleled economic, industrial, and military power. Although not yet in perfect shape, America was reactive and determined to protect the great privileges of its superpower status, along with its prestige and its own social model.

The U.S. counterattack, launched by John Fitzgerald Kennedy when it looked like the Americans would have to settle for the silver medal, ended up reversing the order of finish. Almost at the last minute, the star-spangled astronauts took the top step of the podium, relegating the hammer-and-sickle colleagues to the lower step. The credit goes to the men and to the American system based on the free market, with the connivance of fate which, after having winked at the Reds for quite a few years, supporting their risks and forgiving them for the many, inevitable mistakes, turned its face on them at the most beautiful moment. A betrayal that allowed Neil Armstrong, a Yankee with Scottish and German ancestors, to put his footprint on the lunar soil, the first of the only 12 men, all Americans, who to this day have trodden the dust of our satellite. Since nothing else could be won at that time, the race had to be declared over.

But when did this strange human obsession with visiting the Moon begin? It is always difficult to determine when and how a new page of history, or even a new adventure of knowledge, begins. In general, the events or the personalities that mark a new chapter are the expression of conventions, often arbitrary and more or less imposed by those who, at a given moment, had the right to dictate the rules of the narrative and the interpretation of the facts. As we know, in the aftermath of events, history is usually written by the victors. The conquest of space and the race to reach our satellite are no exception. The beginning of this modern epic goes back through the centuries to the first human being who dreamed of breaking free from the Earth and soaring through the air like a bird.

It is the story of Icarus, the adolescent who ventured to touch the Sun and lost his life for this reckless act. The account of the first aviation disaster was caused by inexperience and youthful daring. Icarus was imprisoned in Crete along with his father Daedalus, a technological genius with poor moral values. The island's ruler, Minos, kept them both in the labyrinth that Daedalus had built for the Minotaur, the monstrous creature born of the union of Minos' wife, Pasiphae, and a white bull. Father and son had been punished[3] for teaching the king's daughter, who was madly in love with Theseus, the woolen thread trick that had allowed the mythological hero to kill the child-eating beast without being trapped in the labyrinth. An intricate tale that combines vices and virtues of men and Gods, free will and fate, in a timeless story that still moves us today for its many psychological insights.

To escape captivity, Daedalus built wings for himself and his son out of feathers held together with wax. When they lifted off, Icarus, forgetting his father's warnings, wanted to fly higher and higher until the heat of the Sun melted the glue. In an instant, exhilaration turned to tragedy. Left without support, the boy fell into the sea. A warning to those who dare too much, out of ambition and thirst for knowledge.

---

[3] According to myth, Minos adopted this restrictive measure in revenge for Daedalus's role in the terrible affair of his wife Pasiphae, mother of the Minotaur, who had fallen in love with a white bull.

The logical step from dream to fantastic narrative is relatively short. Lucian of Samosata, a Syrian of Greek culture who lived in the second century A.D., mocked the genre of the imaginary voyage by recounting *"things that are not seen, nor known by others, that are not and could never be"*. It included an ascent to the Moon on the thrust of the gigantic waves of a storm he had encountered beyond the Pillars of Hercules. This brilliant parody of the Odyssey inaugurated, with a noble pedagogical purpose, the season of fantastic voyages to our satellite, which had remained at the level of a dream with Cicero, and with Dante Alighieri, in the second canto of Paradise, would take on a strongly mystical and scientific character (for what science could mean then).

Ludovico Ariosto, an Italian epic poet of the early sixteenth century, even invented a kind of spaceship, the winged horse with the breast and forelegs of a lion, "which a mare produced from a griffin". The paladin Astolfo used it to restore the sanity of his friend Orlando, who had left him for the Moon because of the unbearable pain of love. A mission impossible,[4] made necessary to return its hero to the Christian army fighting the Moors and thus ensure the victory of the true faith, carried out thanks to a hippogriff, an improbable genetic synthesis of three symbols: agility, strength, and the ability to fly.

About a century later, in order to illustrate his astronomical discoveries from a privileged position in the sky, Johannes Kepler resorted to the expedient of a flight to the Moon that the son of an Icelandic witch, in fact the avatar of the brilliant German scientist, had been able to accomplish with the aid of his mother's satanic arts. Almost contemporaneous with this story, recklessly packaged at the time when Kepler's real mother, Katharina, was about to be imprisoned on charges of witchcraft, there is a novella by Cyrano de Bergerac, a libertine author made famous by the pen of Edmond Rostand. The picturesque character recounts how he tried to reach the Moon using bottles of dew that evaporated upward and finally succeeded due to the thrust of a rocket made of fireworks, set off to celebrate the feast of a saint.

This interesting opening to technology in the context of a mocking boast could have made the big-nosed Gascon in love with the beautiful Roxane the first cosmonaut in history, if only virtually, were it not for the story of a sixteenth-century Chinese Mandarin, Wan Hoo. He had claimed to be able to reach the Moon, also with a chair propelled by about fifty rockets powered by gunpowder. No one knows what happened to this daredevil, although it is not hard to imagine. His disciples, who had acted as firemen by lighting the fuses

---

[4] The hippogriff takes Astolfo to Earthly Paradise where he is welcomed by Saint John the Evangelist who informs him of Orlando's madness and accompanies him to the Moon with a flying chariot.

of the improbable cosmic elevator, reported that the master had disappeared in the smoke of the launch leaving no trace of himself and his primitive spaceship. Sublime madness, real or presumed, has earned the daring Wan a small place in history and another, a little less ephemeral, in lunar toponymy. In fact, his name has been given to a crater located on the hidden face of the satellite, where terrain features abound and many are still unnamed.

In fact, historians tell us that as early as the end of the first millennium, the Chinese were making rockets with black powder. A mixture of sulfur, charcoal, and saltpeter, its properties were accidentally discovered by an obscure alchemist. According to legend, he was searching for the elixir of immortality. Instead, he found an incendiary substance capable of developing a certain amount of heat and gas volume when properly stimulated by fire. The substance was soon used in festivals and wars. Firecrackers to entertain and "fire arrows" to kill: gunpowder-powered rockets attached to a long bamboo stick to give them a minimum of directionality. They were perhaps first used, in vain, by the troops of the Celestial Empire against the Mongols of Genghis Khan's successor during the siege of Kaifeng, the capital of the Jin Dynasty, in 1232.

In the early 1700s, Daniel Defoe, whom we all remember for *The Life and Strange Surprising Adventures of Robinson Crusoe*, took up the theme of a Moon inhabited by intelligent and highly evolved beings. He revealed that he had personally visited this world thanks to a feathered machine, the *Consolidator*, manufactured by the Chinese (whose exotic culture was then very fashionable in Europe) with a technology learned precisely from the inhabitants of the Moon. The British novelist and journalist's intention was to draw the public's attention to the political and social issues of his time, hiding satire behind a bizarre and unreal story in order to exploit its charm. In short, the Moon as a literary expedient.

The work had uncertain fortune. A fate shared by many other writers, before and after, who used the instrument of space travel to achieve other specific goals. For example, to promote and disseminate new astronomical ideas. But no one had yet combined science with imagination to make the fantastic inventions credible, as the philosopher, politician, and Elizabethan essayist Francis Bacon did with his *New Atlantis*, an incomplete utopian novel published posthumously in 1626, dealing with a mysterious "*happy island [that] was known to few, and yet knew most of the nations of the world*".

Jules Verne perfected this approach in his *From the Earth to the Moon*, published in 1865, and in the *Around the Moon* of 1870, which is a continuation of the first novel. Together with other fantastic journeys to the center of the Earth and to the bottom of the oceans, these two stories inaugurated a new and successful literary genre, science fiction, which would have a great influence on the dream and action of the first pioneers of astronautics. The idea with which the novelist from Nantes started out was neither new nor original: to shoot a crew in the direction of the terrestrial satellite with a super cannon.[5]

Eight years earlier, an astronomer and Neapolitan patriot, Ernesto Capocci, had told of a daring mission to the Moon that would take place in the year 2057, exactly two centuries from the time he wrote, with the participation of a woman, Urania. It was a simple allegory of astronomy and not a groundbreaking feminist manifesto, as the main character might suggest. The crew had been transported to the launch site, the crater of an extinct volcano in the Andes, by a balloon named after Giordano Bruno, a symbol of modernity and the break with the shackles of a religion-bound science. Here, a powerful cannon would have given a spacecraft projectile the necessary speed to reach the Moon. Though riddled with naivety and errors, the story attempted to be plausible by paying attention to detail. For example, the use of the recently discovered chloroform to anesthetize the intrepid astronauts and help them endure the trauma of the long journey through the interplanetary void. A precaution that would evolve into the practice of hibernation in the science fiction novels of the second half of the twentieth century.

Written shortly before the dissolution of the Kingdom of the Two Sicilies, Capocci's pamphlet had no significant impact. Few read it, due to the language and limited distribution. The fortune of Verne's stories, told with literary mastery, was quite different. His books were cleverly illustrated and distributed throughout the planet by a skillful publisher. "*We always talk about his unparalleled imagination in predicting scientific inventions. In reality, he was a great reader of scientific journals, which he enriched with what he gradually came to know about ongoing research*", wrote Italo Calvino in the pages of the Italian newspaper *La Repubblica* on January 29, 1978.

The plots of the two French novels follow the structure of the didactic booklet of the Neapolitan astronomer, but the articulation is much more captivating and full of ideas, until the grand finale. The projectile with a human crew fails to hit the Moon and is placed in orbit around the celestial body. The poor astronauts seem doomed to a long agony. However, thanks to a fortunate expedient, the deadly trap is avoided. Fantasy, pathos, and irony

---

[5] Classic launcher used to amaze, as in the story of Baron Munchhausen, or to kill, as in the *Kaiser Wilhelm Geschütz* (Kaiser Wilhelm Gun) and the Hitler's *Schwerer Gustav* (Heavy Gustav), two mega-cannons transported by rail, manufactured by the Krupp steelworks in Essen for each one of the World Wars.

are wisely blended and dressed in reality: the story of a dream that contemporaries, now accustomed to the progress of science and technology proudly displayed in universal exhibitions, could perceive as achievable in the near future.

The impact on the public was even stronger with the serialized novels of English socialist and pacifist Herbert George Wells. *The War of the Worlds* of 1897 and, to stay with the theme, *The First Men on the Moon* of 1901, were enthusiastically received. But neither Verne nor Wells, with their stories, ever crossed the sometimes blurry line that separates imagination, even if prophetic, from rigorous scientific creativity. They did not fully possess the tools to do so. Their voyages to the Moon remained therefore on the level of Greek myth, suggestive, plausible, compelling, and mostly supported by imaginative invention and tempered by common sense.[6]

A quantum leap required a person with a scientific attitude and a heart capable of going beyond the limits of the possible, a competent and ingenious dreamer. This man was born on September 17, 1857—a "*new citizen of the universe*", as he would later say of himself—in the village of Izhevskoe in the Ryazan *Oblast*,[7] a couple of hundred kilometers south of Moscow, with "*one main purpose in life: to do something useful for people*". His name was Konstantin Eduardovich Tsiolkovsky. Today, he is considered the father of theoretical astronautics and rocketry, a science that, prophetically, sent its first cries in Cyrillic letters.

The Tsiolkovsky family belonged to the peripheral middle class of the vast and backward Tsarist empire, constantly balanced between its Asian matrix and its European leanings. Konstantin's parents were both well educated. The mother had enough time to take care of the children. The father, a Pole forced to emigrate to Russia because of his Orthodox faith, struggled to scrape together the bare minimum to get by with his job as a teacher and then as a clerk, not without sacrifice and some hardship. So, the boy's childhood in Ryazan, a city located on the banks of the Oka River in central Russia where the family had moved, passed normally between school, games, swimming in the river, and playing chess with his brothers.

He was just ten years old when things changed radically.

Winter was beginning and I was playing with the sled. I caught a cold that gave me scarlet fever. I was very sick and delirious. Everyone thought I was going to die.

---

[6] Verne, for example, deals correctly with inertia, but not with the absence of gravity to which the occupants of the spherical spaceship would have been exposed. Nor does he address the question of whether the impulse required to travel to the Moon would be tolerable for a human organism, merely imagining a simple fainting spell as a consequence.

[7] Territorial subdivision similar to a region or province.

I recovered, but I had was very deaf. The deafness did not go away. It bothered me a lot.

This was just the beginning of a small private ordeal. The father lost his job, so the family had to move to another city, Vyatka (now Kirov), in the east of the country, along the route of the Trans-Siberian Railway. A cold and hostile place. Deafness prevented Konstantin from attending school and the sleepwalking episodes he already suffered from worsened. In 1869, his mother died and the boy, along with his brothers, was entrusted to a *"semi-illiterate and lazy"* aunt.

These were the *"saddest and darkest"* years of his life, as he later wrote in an autobiography. In the silence of his handicap, he began to cultivate every kind of science on his own, devouring the few books in the home library and constantly mulling over the innumerable doubts of a learning without masters. *"I had no other teacher besides books"*. Frustrations common to many do-it-yourself scientists.[8]

Recognizing his son's great qualities, his father urged him to move to Moscow to expand his educational opportunities in a much more fertile environment. It was a fortunate move, for in the city reborn from the ashes of Napoleonic devastation, the young man, though forced by poverty to skip many meals, finally found a mentor. This is how it happened. Konstantin was supposed to enroll in the Higher Technical Institute, but perhaps because of his relational difficulties, he preferred to do it himself, visiting the great libraries of the city. First of all, the one connected with the Rumyantsev Museum (later the Lenin Library and since 1992 the Russian State Library). An extraordinary structure, created a few years earlier and opened to the public.

At the end of the nineteenth century, Moscow remained a nerve center of Russia, even though it had not been the capital since 1712, when Peter the Great transferred the title to St. Petersburg, on the Baltic coast, with the intention of bringing the country closer to Europe. An argument that Lenin would use in reverse in 1918 to bring the capital back to the banks of the Moskva. Nothing strange, therefore, that Tsar Alexander II, an enlightened despot, authorized in 1862 the move to Moscow of the art collection

---

[8] A shining example of this is Giovanni Battista Odierna, archpriest of the Tomasi fiefdom in Palma di Montechiaro, Sicily, who in the seventeenth century was the first in the world to compile a catalog of celestial nebulae. An ardent student of science, he lamented from the depths of his island that he had no *"socium, vel amicum, aut propinquum, quo paululum sublevari possim. Mens mea praeceptor meus, et difficultates meas nulli communico"* (not a partner or friend or someone close to me on whom I can lean a little. I have only my mind as teacher, and no one else to share my difficulties with). Different places and times, but stories common to "suburban geniuses".

and personal library of Count Nikolai Rumyantsev, who died without heirs, taking it away from St. Petersburg, in order to rebalance the provision of public cultural resources between the two poles.

The vast legacy of works, coins, and books had found hospitality in the Pashkov Palace, a magnificent neoclassical building overlooking the Kremlin, built by a Moscow nobleman and subsequently acquired by the university. The care of the library had been entrusted to Nikolai Fyodorovich Fyodorov, an ascetic figure, founder of the cosmist thought,[9] whose profound influence played a decisive role in Tsiolkovsky's future life and study choices. "*I will help you to study mathematics, and you will help mankind to build rockets, so that at last we will be able to know something more about the Earth, and also to observe it*", the philosopher is said to have written to him one day on a piece of paper. Obviously, Konstantin took the suggestion seriously.

Fyodorov was the illegitimate son of Prince Gagarin and a servant (the intriguing homonymy with history's first cosmonaut, of peasant origin, is purely coincidental). Born in 1829 in his father's country estate, in southwestern Russia, he arrived in Moscow at the end of a hard childhood and a troubled itinerant career as a primary school teacher, complicated by a rigorous application of his utopian and radical ideas. He was convinced that the deficiencies of individuals and human societies were a transitional phase of a path guided by scientific and technological progress aimed at perfection, in an ideal context that contemplated chastity, immortality, rebirth after death, control of natural forces, and the colonization of the oceans and space. Concepts and theses that, with the advent of Bolshevism, pleased the theorists of a new course aimed at the social salvation of all humanity. But the idyll with the Reds did not last long. As the ideological conflict between Soviet communism and religion intensified, the philosopher and his followers were exiled. A *damnatio memoriae* (condemnation of memory) that also affected his tomb where, still in 1928, before Stalin's anticlerical crusade, the inscription "*Xristós voskrés*" (Christ is risen) was present.

An ascetic futurist, this "*brown man of good appearance, of average height, bald but rather well dressed*", lived his own faith, even religious, with rigid consistency. He ate frugally, slept on the floor in a cold cell covered with newspapers, and helped others, both the poorest and the most gifted. To the former, he gave his meager salary and at night he hosted them in the reading room of the library. To the others, including our Konstantin, he offered every

---

[9] Russian cosmism appeared before the October Revolution and developed between the two world wars as a rejection of contemplation in favor of transformation. The goal was to create a new philosophy and, more importantly, a new world, with the end of death, the resurrection of the dead, and the conquest of cosmic space.

possible support for cultural and moral growth. Until almost his death he published nothing, out of modesty and the belief that thought should be entrusted to action rather than to paper and transmitted by example. A man to ridicule or a master to fall in love with. The young Tsiolkovsky fell in love with this "*Socrates of Moscow*", as had already done the giants Lev Tolstoy and Fyodor Dostoevsky. This love changed his life and the history of the conquest of space which, after a long prenatal phase of pure vision, could finally assume the features of a realistic and scientifically founded project.

The two met in the cold halls of the Rumyantsev Library, where the philosopher held public meetings and entertained the young people on various topics of his cosmist faith. Fyodorov immediately grasped the human and intellectual qualities of Kostya, his zeal, acumen, and determination. The boy lived in hardship—"*there was nothing but water and black bread*", he would later recall—using his father's meager allowance to buy books and materials for physics and chemistry experiments. He was always studying, spending what little free time he had in strenuous philosophical discussions with his peers who, unlike him, attended university.

He had opted for chastity, a lifestyle that assured him the possibility of channeling all his energies to learning, without mortifying his passionate soul. He was constantly in love, albeit in a strictly platonic way. An undoubtedly severe approach and yet more humane than that of Christopher Clavius, the stern Jesuit mathematician who, disturbed by the repeated spectacle of a girl drying her long hair in the sun, had the window of his cell at the Roman College, the vehicle of the disturbance, walled up, thus cutting the head off the bull. Konstantin simply separated desire from reality, so that his scientific dream could draw on all his physical and mental resources.

Little is known about the relationship between Fyodorov and Tsiolkovsky, but it was certainly a mutually beneficial partnership. The philosopher saw in the boy's youthful passion for science a means of implementing his ideas: even the most extreme ones such as the conquest of celestial space and the planets, which he conditioned to mankind's ability "*to recreate itself from primordial substances, from atoms and from molecules*", to come to dominate the entire universe. Konstantin found in him the "*university teacher*" he had never had. Thanks to Fyodorov he also obtained a small job in the library, which gave him the material means to continue studying and the opportunity to access all the university texts he had dreamed of consulting. It was in this context that he began to think about space travel, and the cosmos as the future home of mankind. He would never stop. His fertile imagination was catalyzed by the novels of Jules Verne. Fantasy stories that would later claim other important victims.

His stay in Moscow lasted three years. Tsiolkovsky had literally worn himself out with books. Weak in body and with his eyesight so severely damaged that he was disqualified from military service, in 1876 he had to return home because his father could no longer support him. While scraping together some money by giving math and physics lessons, he tried to stay intellectually alive by reading Newton. But life ran hard, gray, and bitter, between hardships and mourning. Finally, after passing the mathematics exam, he found a stable teaching job and enough resources for food, firewood, candles for reading in the long winter nights, and also for a wife. Konstantin had indeed decided to marry the daughter of a priest with whom he was boarding. A girl whom he deeply respected for her culture but did not love, by whom he would have four boys and three girls (five of whom died before the age of thirty). It was a choice in keeping with his determination not to be distracted by the calls of the flesh.

All in all, a seemingly miserable existence that Tsiolkovsky never denied, convinced as he was that poverty, physical handicap, and lack of academic title were the true catalysts of his energies toward the ambitious design of rendering an important service to humanity. A naive utopia? Certainly not, given his character and the other numerous examples of men of genius who have implicitly, and sometimes openly, claimed the maieutic value of life's difficulties. For example, the Turin-born Joseph-Louis Lagrange, prince of mathematicians, attributed his decision to devote himself to science to the wickedness of his father, who had driven the family to the pavement with gambling. If he had remained wealthy, he thought, he might have settled down on the soft cushions of a golden existence, without seeking the "cheaper" pleasures of the intellect.

For Konstantin, the time had come to apply to concrete problems the wealth of scientific knowledge he had painstakingly acquired. The publication of his first works brought him a breath of fame and a place in the scientific community. He thought, acted, and dreamed. He began to speculate about the physical behaviors of human beings and things in a space devoid of atmosphere and external forces. He built models for his students, including rudimentary aerostats, and when the deafening noise and the care of his first children allowed him, he wrote astronomical fantasy stories.

In the meantime, he had moved to the nearby Kaluga, the capital of a region that had just been sculpted by ancient glaciers, cold as befits a place far from the sea, where he would remain from 1892 until his death. It was here that he wrote his first scientific papers on space travel and the problems of using human crews. The text was full of calculations, demonstrations, assessments, and revolutionary proposals: for example, that of a liquid propellant made of a mixture of hydrogen and oxygen, which took science fiction into the dimension of the possible. He also began to take an interest in philosophy. A sign of a new maturity that allowed him to lift the veto against purely abstract activities that he had imposed on himself during his stay in Moscow.

Then came 1903, the *annus mirabilis* for human flight in the blue skies and in the darkness of deep space. On December 17th, at Kill Devil Hills in North Carolina, Orville Wright took advantage of a strong headwind to lift off from the ground Flyer 1, a rudimentary airplane he had built with his brother Wilbur. It was the first time for a powered object heavier than air. The thirty-six meters covered in twelve seconds at grass level by a bicycle manufacturer lying along the horizontal structure to control the tail of an improbable aircraft marked the birth of the aviation era. A long chased mirage that materialized with difficulty which in just a decade would become a bloody reality in the skies of the Great War and then the main vehicle to physically connect the most remote corners of the planet.

In this same year, Konstantin Tsiolkovsky published his most important scientific work entitled, *The Exploration of Cosmic Space by Means of Reaction Devices*. According to historians of cosmonautics, it marked the turning point between the season of fantasy and that of space exploration. The document was written with a heavy heart due to the suicide, the previous year, of his nineteen-year-old son, who poisoned himself because he could not bear the endemic misery.

Russia, too, was in deep crisis. Backward, held back in its capitalist development by centuries of privilege, it was shaken by the quivers of a dissatisfied population. The peasants demanded greater social justice, as did the urbanized manufacturing workers, who were beginning to organize themselves into factory councils called "*Soviets*". The intellectual bourgeoisie and liberals were also restless, and the military was seething, humiliated by the disastrous defeat in the war with Japan for control of Manchuria and Korea. In this context, a spark struck during a peaceful workers' demonstration in St. Petersburg in December 1905 was enough to ignite a chain revolution. It was the first act of a long drama that would lead to the downfall and death of the Romanovs and the transformation of Tsarist Russia into the Union of Soviet Socialist Republics (USSR), in those ten days of November[10] 1917 that "*changed the world*".

In 1903, the year of Wright's flight and the publication of Tsiolkovsky's treatise, insurrection was still far. The temporal coincidence of the two facts,

---

[10] The Russian Revolution of 1917 is called the October Revolution because according to the Julian calendar, which was still in use in the Russian Empire, it conventionally began on October 25, a day that is instead of November 5 in the Gregorian calendar.

both inherent in the same kind of dream that led Icarus to his death, is indeed intriguing as it marks also a common starting point for the two leading nations in the space race. But the analogy ends there. In fact, airplanes need an atmosphere to fly, which is used for propulsion and lift. Rockets must do without it, and this fact implies some similarities and many substantial differences.

Observing the birds that sometimes soar and circle in the sky without flapping their wings, humans have learned the trick of "swimming" in the air despite their weight. The solution rests on a more general effect studied by the Swiss mathematician and physicist Daniel Bernoulli in the eighteenth century. The idea is to design two fins (the wings) in such a way that the flow they encounter as they race through the air is forced to follow paths of different lengths on the two sides: longer, and therefore faster, on the humped back compared to the flat underside. This simple geometric-dynamic trick creates an upward pressure that counteracts gravity and manages to support a body heavier than air,[11] provided that the speed and density of the fluid are sufficiently high (relative to the mass to be lifted).

To make an airplane move—a necessary but not sufficient condition for flight—you need only screw a propeller into the air or feed atmospheric oxygen into a jet engine, if you want to use the same physical principle that applies to rockets. But these latter, whether ballistic missiles or launchers of space vehicles, must travel in an environment where the air is very thin or totally absent. This circumstance conditions both the lift and the supply of the oxidizer necessary to activate the chemical reactions in the jet engine. The problem is serious but not insoluble.

Let's start with the issue of thrust, which is solved in the same way as jet engines, by invoking the third law of dynamics. In 1687, in modeling the effects of forces on bodies, Isaac Newton postulated that for every action there must be an equal and opposite reaction. The idea is as follows. When body A experiences the action of a force from body B, the latter in turn experiences a force from A of equal intensity and directed in the opposite direction. In other words, what is done is returned! Demonstrations of this phenomenon in everyday life are not lacking. For example, the recoil of a gun at the moment of firing is nothing more than the reaction of the weapon to the action of the gases produced by the explosion on the bullet.

To understand how the third principle can be useful for the applications we're interested in, let's do a little thought experiment of the sort that Albert

---

[11] The lighter ones, such as balloons and dirigibles, simply use Archimedes' thrust, that is, the excess of upward pressure that keeps the mass of air in perfect equilibrium with gravity within the volume occupied by the flying object.

Einstein was so fond of. Imagine that you are standing in the middle of a pond whose ice surface has no friction whatsoever. In these conditions (ideal for the example, but not for you!), you will not be able to reach the shore because, according to the hypothesis we have made, neither the soles of your shoes, nor your gloves, nor any other part of you and your clothes have any grip. At most, you could rotate around your center of gravity, but you could not translate because, in the words of Archimedes, you lack a "fixed point". It's a serious problem. It's very cold and night is approaching. You risk freezing. How do you get out of this predicament? By applying a little elementary physics. For example, all you have to do is take off your shoe and throw it away with as much energy as possible. As a reaction you will slide in the opposite direction and with a little patience you will reach the shore and safety. You would do it faster if instead of the shoe, which weighs little, you could throw something heavier, like a bowling ball or the large volume of a Greek-Latin dictionary that you have for some reason brought with you.[12]

A jet engine works in the same way. It turns chemical energy (part of what atoms retain in their electronic shells) into thermal energy (chaotic motion of gas particles) by oxidizing some kind of fuel, such as alcohol, and then converts this heat into mechanical energy by letting the pressurized gas flow through one or more nozzles. The motion is generated, as in the case of a balloon properly inflated and then pierced with a needle, by the dynamic reaction to the ejected gas, which increases with the mass of the jet and its velocity.

This all seems clear and simple, but it is not. To understand where the catch is, let's go back to the frozen pond experiment. In order to gain speed, you had to sacrifice a shoe and remain half barefoot. Even the balloon that flies when you puncture it, arrives deflated. In short, to create a reaction, one must break an entity into two or more parts that are not necessarily equal. The jet engine gradually loses its own fuel. Airplanes constantly renew the oxidizer by sucking in air from the surrounding environment and then compressing it with a turbine. This is not possible with rockets. In order to function, these flying machines must carry everything they need to produce the chemical reaction that propels them, and they must accept the gradual loss of this fuel, which is ejected at high speed during the motion. In this way, their mass changes over time, like a cart full of sand with a hole in the bottom of the loading platform. The consequence is not trivial.

---

[12] In fact, the principle of action and reaction implies that the center of gravity of the system that splits (you and your shoe in the example), under the impulse of an internal force only (that of the arm muscles), remains stationary and the two parts separate, moving in opposite directions with speeds inversely proportional to their relative masses.

It was precisely Tsiolkovsky, lost in the boundless Russian countryside, who analyzed the problem mathematically, proposing in the fundamental article of 1903 an equation that now bears his name and that he had already presented six years earlier in a work with an unmistakable title, *Rocket*. The formula, which Konstantin worked out within the framework of Newtonian mechanics, tells us that the thrust produced by the engine is proportional both to the speed at which the burned gases flow and to the relative variation of the residual mass, which is very small at the start and increases as the tanks are emptied. If you think about it, the same thing happens with Formula 1 cars, which go faster at the end of the race (if the tires hold up) because they are lighter.

Tsiolkovsky was able to predict, with his research, what would actually happen in the skies over England in late 1944 with Wernher von Braun's flying bombs, and in the stratosphere with the *Sputnik* in 1957. He firmly believed that "*what is impossible today, will be possible tomorrow*", and with this belief he imagined and described the solution to countless problems that he foresaw: brilliant ideas about guidance, propulsion, and rocket management. But no experiments, however. It would not have been possible. There was a total lack of technology, industrial and economic resources, and of context.

However, he did not defy the vision, as evidenced by this type of reasoned prophecy:

> *There was a time, very recent, when the idea of the possibility of knowing the structure of the heavenly bodies was considered bold even by the most renowned scientists and thinkers.*[13] *That time has passed. The idea of a more direct and*

---

[13] In 1835, in a work devoted to the philosophy of the physical sciences (*Cours de la Philosophie Positive*), the sociologist Auguste Comte wrote:

> *On the subject of stars, all investigations which are not ultimately reducible to simple visual observations are [...] necessarily denied to us. While we can conceive of the possibility of determining their shapes, their sizes, and their motions, we shall never be able by any means to study their chemical composition or their mineralogical structure [...] Our knowledge concerning their gaseous envelopes is necessarily limited to their existence, size [...] and refractive power, we shall not at all be able to determine their chemical composition or even their density [...] I regard any notion concerning the true mean temperature of the various stars as forever denied to us.*

Two reasons for this peremptory statement: the impossibility of making the necessary measurements both at a distance and in situ, by means of space travel. Two blatant errors in one stroke by a man of genius but without vision! A similar short-sightedness was shown by Lord William Thomson, first Baron Kelvin, when he declared at the beginning of the twentieth century that there was nothing left to discover in physics. All that remained to be done was to complete the knowledge that had been acquired. An entirely different point of view from that of Enrico Fermi, who held that "*if the result confirms the hypothesis, then you've made a measurement. If the result is contrary to the hypothesis, then you've made a discovery*".

closer knowledge of the universe appears to us today, I believe, even more bizarre. To set foot on the soil of asteroids, to hold a lunar rock in the hand, to build mobile stations in space, to form living rings around the Moon, the Earth, the Sun, to observe Mars from a few tens of kilometers, to go on its satellites or even on its surface.

He was right. Today we have already achieved all this, and even more, thanks to the use of "*reaction devices*" (rockets). We have walked on the Moon several times, and we plan to do so soon on the surface of Mars, where roving laboratory rovers are currently roaming as our eyes, hands, and outposts on the Red Planet. We have landed on asteroids scarred by craters, wounds from ancient Star Wars when the Solar System was still teeming with the remnants of the construction site, and on the tiny nucleus of a comet about half a billion kilometers from Earth, no bigger than a village and unable to offer the guest a suitable gravitational anchorage. We travel beyond the Pillars of Hercules of the planet Neptune, toward the icy and dark world of the dwarf planets, and down into the reserve of solar comets. Thanks to the taxi service offered by the Russian *Sojuz* spacecraft, we have maintained a human outpost 400 km high in a space station, the ISS, which has been an icon of scientific and technological progress and a positive example of cooperation among all the peoples of the Earth, or almost. An example that, like so many others in the long history of mankind, has been squandered by man's inability to live in peace.

Many of the solutions and precautions that have made this latest glorious season of the space epic possible were evaluated and proposed by the master of Kaluga: compensation chambers, double walls to defend against meteorites, combustion chambers, space stations, and multi-stage rockets which he called "*cosmic rocket trains*".

Konstantin Tsiolkovsky died on September 19, 1935. He had just turned 78 years old. A week before he closed his eyes, he wrote a letter to Stalin to reaffirm his faith in the principles of the October Revolution and in the person of the leader. Now old and tired, he wanted to sum up a life spent in the service of humanity, just as his old teacher Fyodorov had asked him to do. He felt that he owed a debt of gratitude to the "*party of Lenin and Stalin, of the Bolsheviks and the Soviet power, authentic leaders of the progress of human culture*", for recognizing his efforts. He therefore intended to donate all his works "*on aviation, astronomy, and interplanetary travel*" to his homeland, confident that someone would be appointed to carry out his mission successfully. He was wrong, but only temporarily.

In the years to come, Stalin would be interested in something else: in the great transformation of a predominantly agrarian society into an industrial power; in the great purges of his and the people's enemies, real or presumed; in leading the heroic *Velikaja Otečestvennaja Vojna*, the Great Patriotic War against Nazi Germany; and then in the struggle for world domination with the hated Western capitalists, igniting the flames of socialism where the grass of life was driest, while isolating his empire behind the insurmountable walls of the Iron Curtain. It was only after the war that he developed a lukewarm interest for missiles as strategic weapons against the nuclear power of the United States.

Tsiolkovsky had not taken an active part in the revolutionary movements that brought the Bolsheviks to power, but he fully shared the ideals of Marxist communism and had positively judged the end of a war declared by the Tsar but hated by the people. "*The USSR is treading the great path of communism and the industrialization of the country with strength and perseverance, and I cannot but support it*", he would write. However, if the old rulers of Russia had mostly ignored him, the new masters did not behave much better, only offering him a nomination to the Soviet Academy in 1918, which Konstantin could not accept, as it provided for the transfer to Moscow.

Now retired, he remained in his Kaluga to study until, in 1923, the publication of the doctoral thesis of the German Hermann Oberth, entitled *The rocket into planetary space*, did not open the eyes of the Soviet authorities, who rediscovered their champion, quickly forgetting to have previously interrogated him in the Moscow offices of the secret police for some writings judged anti-revolutionary. Russian roots were needed for every branch of science, technology, and culture in order to create the glue of memory around the new regime. The elderly Konstantin fitted perfectly into the role of the father of cosmonautics.

It was a temporary role, however, because in the twenties only a handful of adventurous pioneers really believed in the scientific and military importance of rockets. In the next two chapters, we will look at three of these pioneers, one American and two Germans, to tell the story of how the space race moved from words to facts.

An Introverted Genius

When you walk on the Earth after flying, you will look at the sky because there you have been and there you will want to return.
Leonardo da Vinci
What you can do, or dream you can, begin it. Boldness has genius, power, and magic in it. Begin it now
Wolfgang Goethe

The dream of touching the Moon began to take shape in the 1920s, when the rockets the Chinese had been playing with for so long became the prototypes of innovative flying machines. At the dawn of the new century, the solitary Tsiolkovsky had dictated the rules of the game and showed the way by mixing science and fantasy together, without falling into science fiction. The path had been charted. Now a pretext had to be found to transform a scientific curiosity into a necessity, in order to gather the human and financial resources indispensable to put the theory into practice.

This fundamental and daring step was taken in the wake of a very brutal world war and an unjust peace, inspired by hatred and fear. It involved, for different reasons but with identical passion, both the winners and the losers. The main actors were an American physicist, Robert Goddard, and two German engineers, Hermann Oberth and Wernher von Braun. The hand, played in Kaluga with vehement creativity by Tsiolkovsky, had passed, albeit temporarily, to other players under other flags.

© The Author(s), under exclusive license to Springer Nature
Switzerland AG 2024
M. Capaccioli, *Red Moon*, Springer Praxis Books,
https://doi.org/10.1007/978-3-031-54760-7_3

The Russians,[1] in fact, had other things to think about. The great empire of the Romanovs had collapsed catastrophically under the pressure of internal tensions and serious failures in the war against the Black Eagles of Austria-Hungary and Prussia. The anti-system revolution that exploded in 1905 had never died out. The openings to greater democracy made by the weak Tsar Nicholas II and even the harsh repression by his guards were not enough to quell it.

The intellectuals had raised their heads, even ready to have them cut off for a good cause; the bourgeoisie, excited by the confrontation with the West, was elbowing its way forward; and a people of servants had begun to demand more humane living conditions and greater social equity. An indigenous awakening of consciences, inspired by the winds of freedom that had laboriously arrived from Europe, which the St. Petersburg court, blinded by centuries of absolute power, was unable to perceive.

The house of cards was in the balance, supported by the power of tradition and the fear of the new. To make it collapse completely, a push was needed. A war that the Tsar did not want to declare and did not know how to conduct provided it. In February 1917, military disasters and serious territorial losses led even the most loyal conservatives to demand the deposition and arrest of the sovereign and the establishment of a provisional government to face the emergency of a shattered and exhausted country.

The Prussians, who had defeated the Russian army in two major battles and wanted to withdraw their large forces deployed on the eastern front to concentrate them against the British and French, also gave economic aid to the rebels. Returning to his homeland from exile with the complicity of the enemy, Vladimir Lenin took over the leadership of the popular movement. It was necessary to bring order to the archipelago of soviets, torn apart by the struggle between the moderate wing and the Bolsheviks, and to form a common front to repel the counterrevolutionary uprisings. At dawn on October 25, Julian Calendar, the workers' militia stormed the seat of the democratic government in Petrograd.[2]

Thus began a season of history in which the social and economic theories of Karl Marx and Friedrich Engels were applied for the first time. "*We have*

---

[1] The adjective Russian is used here instead of Soviet. In the Russian language, confusion is avoided by the existence of two forms, one for ethnicity and the other for nationality. The Russians (*Russkij*) are the majority ethnic group in Russia, where today they make up 80% of the total population with almost 116 million individuals. *Rossijane* (plural) is the collective name for all citizens of Russia, regardless of their ethnic origin. The term has been used since the beginning of the sixteenth century. In the 1990s it received its modern meaning, that of citizens of the Russian Federation regardless of nationality.

[2] This is how St. Petersburg was called from 1914 to 1924 by the will of Nicholas II, who considered the previous name "too German"; then it became Leningrad until 1991.

now raised the white flag of surrender", Lenin is reported to have said at the signing of the Brest-Litovsk Peace Treaty with the Prussians, on March 3, 1918. "*Later, we will raise the red flag of our revolution over the whole world*". The rest of the twentieth century witnessed attempts to fulfill this promise. The experiment, which cost fear, tears, and blood, ultimately failed, both in Great Russia and almost everywhere else on the planet where it was tried. Like many utopias, the communist ideal had to confront the harsh edges of reality and the immaturity and selfishness of human beings.

It left us with much pain, some remorse, a great disappointment, but also some masterful insights in many and various fields, also the result of confrontation with the capitalist and liberal West. Among these the conquest of space stands out, initiated for reasons quite different from the designs and projects of the master of Kaluga. Tsiolkovsky's dream could not interest Lenin, who was entirely focused on creating a nation out of chaos, or his successor Joseph Stalin, who wanted to rebuild the old empire with new social rules under the banner of the hammer and sickle.

Born of a bloodless revolution, the new Russia had to face the hostility of the world from the outset, embodied by the reaction of the Tsar's old allies. England and France, joined by the United States and the loyalist groups of neighboring nations, believed that Russian soldiers had to be brought back to fight so as not to give the Prussians an advantage. The Westerners also wanted to teach the rebellious working class a resounding lesson, to prevent the "pernicious" desire for social justice from creeping out of the country's borders and contaminating the world, just as Lenin had promised. It was a civil war that also cost the imperial family their lives, massacred *en masse* to prevent them from leading the nostalgics.

The fratricidal struggle lasted until 1920. The Bolsheviks won, and Lenin ruled the country until 1924, when he died of a cerebral hemorrhage after a long illness. Having put an end to the "war communism", in a short time he had forged the tools of propaganda, persuasion, and repression that for seventy years would characterize the new great nation. Powerful and well-lubricated weapons that will also have consequences in our history. He was succeeded at the head of the newborn Soviet Union[3] by Ioseb Besarionis dze Jughashvili, better known Comrade Stalin, whom he himself had promoted

---

[3] The command hierarchy of the Union of Soviet Socialist Republics provided for an actual leader in the person of the General Secretary of the CPSU (Communist Party of the Soviet Union), elected by a Central Committee. This latter also appointed the members of the Politburo, called the Presidium from 1952 to 1966; a Chairman of the Presidium of the Supreme Soviet (legislative body) with the nominal function of head of state; a Chairman of the Council of People's Commissars (administrative body); a Chairman of the Council of Ministers (executive body).

to General Secretary of the single party, even though he liked him less than the moderate Lev Trotsky.

In the United States, emerged stronger than before from the Great War, Robert Goddard (1882–1945) was thinking about liquid-fuel rockets: a technology anticipated in its essential lines by the reflections of Tsiolkovsky and then considered in Germany by Oberth (1894–1989), proving that, normally, good ideas do indeed spring from a flash of genius, but in a context now mature to let them blossom. In short, nothing comes from nothing, not even revolutions. The subject of propellants is so important that it deserves a digression.

Gunpowder rockets first appeared in China in the thirteenth century, during the Song Dynasty's struggle against Mongol invaders. They were still powered by the same fuel in modern times when their military application emerged. The British had discovered them with this function in the late eighteenth century while fighting the Mysore Tiger in southern India. Sultan Fateh Ali Sahab Tipu, a very cultured, refined, and imaginative monarch, had formed a corps of five thousand gunners specialized in the management of metal rockets armed with frontal lances with a range of about 2 km. These exotic *Wunderwaffen* were not enough to defeat the sturdy soldiers of His British Majesty, but they caught the attention of the English colonel William Congreve. He studied and perfected them to the point where the rockets became weapons worthy of a large modern army (and an excellent source of income for himself).

Clad in iron and with a range of up to 3 km, the Congreve rockets were used by the British against Napoleon's troops in two fierce attacks at Boulogne and Copenhagen. With a payload consisting of a 3 kg incendiary bomb, they were highly appreciated by the Navy. Their trajectory was controlled by a long tail, a wooden rod 4 m or more in length, which limited the rocket's wobble after launch: a trick known to the Chinese for six centuries, based on a relatively simple physical principle.

In passive control systems, i.e., without any form of intelligence to make the appropriate corrections to maintain the course, the center of thrust— in our case the rocket nozzle from which the exhaust gas escapes—must be located in front of the center of mass (relative to the direction of motion). This configuration offers some stability, as anyone who drives a rear-wheel-drive car, sportier but more prone to spin, knows from experience. This is also why we find it so tiring to send the supermarket cart straight, and why we pull the cart instead of pushing it.

Congreve's rod was used to shift the center of mass of the system so that it felt behind the rocket's nozzle. The gimmick worked reasonably well,[4] but it was cumbersome. In the mid-nineteenth century, another Englishman, William Hale (1797–1870), found a way to replace the long tail with a rapid rotation of the rocket around the axis of motion, achieved by the twisting action of pairs of nozzles oriented in opposite directions. Nothing new. It was an active version of a technological gimmick long used[5] in firearms, achieved by a helical rifling made inside the barrel. The rotation imposed by this geometry on the firing gases was transmitted to the projectile in order to maintain its orientation and thus facilitate its penetration in the air and in the target. A magic that finds its explanation in the so-called gyroscopic effect. Don't panic! As children, we all played with this physical toy. We simply called it a top.

In general, a gyroscope is any solid body whose mass is symmetrically distributed about an axis around which it freely rotates. The laws of mechanics require that such an object stubbornly maintains the orientation of the axis (relative to an absolute reference system set by the "fixed stars") as long as no external forces intervene to disturb it.[6] This ideal isolation is achieved, for example, by mounting a flywheel on a gimbal with virtually frictionless joints. In short, the gyroscope is like a finger firmly pointed at a hypothetical star that never loses its grip and can therefore be used as a reference in space. This property makes it an almost ideal tool for monitoring changes in direction and if some kind of rudder is provided, for controlling it in order to correct the trajectory of our rockets.

More simply, the gyroscopic effect can be used to keep a projectile in a chosen direction. It will be enough, as we said above, to make it rotate around the axis of symmetry, as in the Hale missiles. The latter entered the arsenal of the U.S. Army, which used them against the Mexicans in the war sparked in 1846 by the question of sovereignty over Texas. Later, a third Englishman, Colonel Edward Boxer (1822–1898), devised a stratagem to extend the range of these weapons and created the first two-stage rocket, which also found some use in the Navy in peacetime.

However, the technology of solid-fuel rockets did not seem to offer performances comparable to those of modern artillery, except for the reduced size of the launchers. For this reason, it was gradually abandoned by the military,

---

[4] For the rod to be effective, the rocket had to be in motion. Therefore, it had to be guided by a "launch channel" at the start.

[5] Early attempts at rifling musket barrels date back to gunsmiths active in Vienna and Nuremberg in the late fifteenth century.

[6] This stubborn physical entity reacts to the disturbance with a conical motion (of precession) around the original direction of rotation.

which only rediscovered it with the bazooka, the anti-tank rocket launcher invented in 1942 and widely used by the Americans in World War II and in the USSR with the Katyusha multiple rocket launchers. As military interest waned, so did the motivation and resources to develop alternative technologies. And no one really thought that a weak and uncontrollable thrust would be enough to get into space, at least in the short term. Gunpowder wasted much of its energy in producing large particles (smoke) that were useless for propulsion, and once ignited it could not be controlled. It burned progressively on its own until it was completely consumed, like a candle that cannot be blown out to be reused.

That's why the most creative minds began to think about the possible use of liquid fuels. For example, hydrogen and oxygen, which, when combined into water molecules, yielded a relatively large amount of clean energy,[7] the delivery of which could be modulated by a simple tap. But why liquids and not just gas? To reduce volume so they can be stored in tanks of limited size. For example, liquefied oxygen takes up 860 times less space than it needs to expand to gas at 20 °C.

Seen in this light, the problem of building a liquid-fuel engine seems relatively simple. Two tanks, one for the material to be burned and another for the oxidizer, a system for pumping the liquids into a combustion chamber, one or more nozzles to expel the combusted gas under pressure from which the thrust[8] is derived, and that's it! In practice, however, the difficulties are numerous, starting with the cooling of the combustion chamber, where the temperature reaches 3,000 degrees. Then, the nozzles must be designed to maximize their efficiency, since the power available is still limited compared to the needs and cannot be wasted. And once thrust is achieved, the capricious thoroughbred must somehow be guided along the desired trajectory. None of this had ever been attempted until one freezing afternoon in mid-March 1926.

The scene is a snow-covered field on Pakachoag Hill, near the village of Auburn, Massachusetts. The land was part of the farm of an aunt of Robert Goddard. Here a metal tripod similar to a clothesline or the frames that hold children's swings had been fixed to the ground. It served as a support for an object resembling a postmodern sculpture, consisting of three tanks joined together and connected to a traditionally shaped rocket: a pointed cylinder with a conspicuous nozzle on the back.

---

[7] A concept that is driving fuel cell technology.

[8] Many, blatantly wrongly, believed that this thrust had to be directed against something, and therefore excluded the possibility that rockets could work in a vacuum.

Overall, the project followed the geometry of the Congreve rocket, with the guide rod replaced by the three tanks, each containing gasoline, liquid oxygen, and an inert gas under pressure. The fluids communicated with the combustion chamber through long aluminum connecting tubes. In the absence of suitable pumps for this purpose, Goddard had devised a pressure-assisted delivery system, provided directly by evaporation in the case of oxygen, and by the inert gas introduced into the tank in the case of gasoline. The cryogenic oxygen also served to cool the casing of the combustion chamber, which was exposed to extreme temperatures. Finally, a funnel-shaped shield protected the tanks from the rocket's fiery exhaust, which towered over this unlikely structure.

Bundled up in a long coat, with gloves, beret, and military boots to protect himself from the biting cold, Goddard led a meager launch team consisting of two assistants from the local university where he taught physics. His wife, a staff member in the provost's office whom Robert had married two years earlier, was also present. Armed with a camera, she was ready to immortalize the event, but at the most crucial moment, she ran out of film.

At 2:30 p.m. the engine fired, but for the next twenty seconds, the rocket refused to move. There was not enough thrust to counteract its weight. Then, after dumping a good portion of its fuel, the bizarre ancestor of modern spacecraft managed to lift off. Its flight lasted only two seconds due to the rupture of a nozzle that caused it to swerve. It reached an altitude of 12.5 m and covered a distance of 56 m at a speed of one hundred kilometers per hour, after which it crashed ingloriously on the frozen ground of a cabbage field. Today, the launch site is a National Historic Landmark protected by the United States government.

It would seem to be the story of a comic disaster. Instead, it was a resounding success. Goddard had demonstrated that rocket propulsion was possible, even with liquid propellant. This news opened the gates of paradise to those who had dreamed of conquering space but feared having to accept the inefficiency of solid fuels. It also, as we shall see, gave arms manufacturers an opportunity to reconsider the military virtues of rockets. But who was the architect of this revolution?

Born in Worcester, Massachusetts, in 1882, Robert Goddard came from an old New England family. His father worked a variety of jobs and in his spare time dabbled in simple physics and chemistry experiments. It was he who instilled in his only son his first scientific curiosities: the production of static electricity with all its bizarre consequences, observing the sky with an amateur telescope and studying insects under a microscope. He also gave him a subscription to *Scientific American* magazine.

In Worcester, the context was that of classic rural America. Wide open spaces barely interrupted by rolling hills, clean air, the smell of grass, contact with animals and inanimate nature. An ideal place to grow up. And yet, shortly after his birth, little Robert found himself in Boston, where the family had moved. A hell of a few dozen miles from paradise. He did not like the big city. He suffered from stomach and respiratory ailments, aggravated by smog and crowds. His fragile health prevented him from attending classes regularly and would later affect his character and professional life. He lost a few years of schooling, but not the opportunity to learn. He read about science by plundering the city library and was constantly fantasizing. And when his father took him to relatives in Worcester for the holidays, he enjoyed the countryside, flying kites, walking in the clearings and woods, climbing small heights, and practicing with a rifle. A boy like many in a large and young country, with his roots firmly planted in the Earth. Meanwhile, his mother had contracted tuberculosis. The Goddards had to return permanently to Worcester in search of healthier air.

Robert was 17 years old when, to his recollection, he first conceived the desire to fly in space. He had climbed a neighbor's cherry tree and, while admiring the panorama of the fields disappearing on the horizon, he imagined *"how wonderful it would be to make some device which had even the possibility of ascending to Mars"*. *"I was a different boy when I descended the tree from when I ascended"*, he would later say, because *"existence at last seemed very purposive"*. Memories embellished to fit the biography of a genius? Probably not, because for the rest of his life, Goddard regarded that day as an anniversary to be celebrated every year.

Around the same time, he came across *The War of the Worlds,* the science fiction novel by H.G. Wells. A real spark for the imagination of a young, down-to-Earth visionary. *"The dream of yesterday is the hope of today and the reality of tomorrow"*, he would say in 1904, in a kind of programmatic testament contained in a speech for his high school graduation ceremony. His dream was to navigate in space and as a simple student he told it in an article for a popular science magazine, which however rejected it. The world was not ready.

As he grew older, Robert began to perform more complex experiments than the simple kites he had enjoyed as a child. For example, he designed and built an aluminum balloon in his home garage and filled it with hydrogen. He claimed it could fly, but the sphere was too heavy and stayed nailed to the ground. It was a bitter disappointment that the young man diligently recorded in a systematically updated diary, in which, in addition to his emotions and life experiences, he concisely reported all the notes related to his experiments. He never discontinued this practice.

Eventually, his health improved somewhat, allowing him to graduate from high school with honors. He enrolled at the Worcester Polytechnic Institute and in 1908 received a bachelor's degree in physics. He had chosen this subject because he had convinced himself, after reasoning and some elementary tests, that "*if a way to navigate space were to be discovered, or invented, it would be the result of a knowledge of physics and mathematics*". With that thought in mind, he taught freshman physics for a year to make ends meet. Then he enrolled at the University of Worcester: a young college, founded only twenty years earlier thanks to a bequest from a wealthy businessman. Jonas Gilman Clark had amassed a large fortune through real estate speculation in California and, at the suggestion of Leland and Jane Stanford, his friends and benefactors of the famous Palo Alto University, had invested part of his wealth to promote the higher education for his fellow citizens. A patronage almost unknown in Europe today. Despite its young age, Clark University had a distinguished educational pedigree. Albert Michelson, the first American to win the Nobel Prize in Physics in 1907, had taught there, and there the legendary Sigmund Freud had given a series of famous lectures, which Robert himself had probably attended.

In 1912, after quickly earning both a master's degree and a doctorate, he moved to Princeton to work at the Palmer Physics Laboratory on a fellowship. But after only a year, he had to return home for serious health reasons. Like his mother, he had contracted tuberculosis. An ordeal. It was even feared that he might die. But the young man was not ready to leave. He had a dream in suspense and made every effort to recover. He succeeded and at the same time developed a kind of pathological distrust of others. He had to protect his ideas and inventions at all costs, using the tool of the patent almost compulsively. By the end of his life, he would have accumulated over 200.

After recovering, he returned to Clark as a part-time researcher in the fall of 1914. War had broken out in Europe, reluctantly declared by the Tsar and strongly desired by the Kaiser. The Prussians had launched the first lightning attack on France, evaded its defenses by invading neutral Belgium, and committed atrocities, real or alleged, that aroused the indignation of the world. In response, and in order to defend the reasons for their patriotic commitment, the leading German scientists signed and circulated a manifesto on October 4, 1914, in which they promised to "*carry on this war to the end as a civilized nation, to whom the legacy of a Goethe, a Beethoven, and a Kant is just as sacred as its own hearths and homes*". They were not true to their word, judging by their use of deadly and insidious weapons such as nerve gas: a crime readily imitated by their opponents. By serving death, science had begun to "*know sin*", to quote Robert Oppenheimer's sad remark after the atomic bombing of Japan, establishing a macabre custom that would be revived twenty years later with the discovery of the offensive potential of the atom and modern missiles.

Once the initial momentum was exhausted, the bloody melee had turned into a war of position. A deadly drip of pain and blood for Europe's tragic farewell to world leadership. America had temporarily withdrawn from the drama, entrenched behind the principles of the Monroe Doctrine, which can be summed up in a cynical, often useless "*let's just mind our own business*".

In the quiet of Worcester, Goddard was planning the work of a lifetime. He had received permission from his university to work on rocket propulsion: a fundamentally new discipline for which there were some brilliant proposals, elegant calculations, but very little quantitative information, starting with a reliable measure of engine efficiency. For this purpose, Robert built pendulums and then spring systems to be able to estimate from their reaction the fraction of kinetic energy per unit of chemical energy extracted from the solid propellant. Laboratory tests using scale models of rocket engines showed that the efficiency of available solid propellants was almost ridiculous: 98% of the chemical energy, and thus of the passive weight of the fuel, was lost. As Tsiolkovsky's rocket equation clearly showed, it was necessary to better exploit the gas thrust by increasing its ejection velocity (relative to the rocket). This also meant changing the nozzle geometry. But how?

While scanning the literature, Robert came across a project developed by a Swedish engineer of French origin to improve the efficiency of steam turbines. Instead of the traditional converging nozzle, in the late nineteenth century, Gustav De Laval had patented one in the shape of an inverted funnel. The physical trick was to use a constriction at the point where the cup was attached to the nozzle to accelerate the flow of hot gas to the speed of sound (relative to the rocket), and then let it expand in the terminal divergent part to reach supersonic speed. It was exactly what was needed. With the De Laval nozzle, static efficiency increased up to thirty times and speeds reached two thousand meters per second. Much better than steam engines and even the internal combustion engine patented by Rudolf Diesel in 1892. Such a powerful thrust justified the hope of carrying a vehicle into the empty space that separates the Earth from the Moon.

Goddard firmly believed it could be done, but his confidence went against the grain. Conventional wisdom held that the fluid jet needed an obstacle to act as a propulsive force. In short, "no air, no motion". For this reason, he pragmatically decided to promote his work by focusing on the benefits that reaching the top of the atmosphere (where air is scarce but still present) would bring to meteorology and the Earth sciences in general. He needed money and support. Being the practical man that he was, he added a dash of realism to his dreams.

Meanwhile, despite President Woodrow Wilson's firm opposition, the United States was sliding toward an armed commitment against the Central Powers. More than the ideal choice of side, it was a matter of protecting the commercial interests of American companies and the colossal loans granted by U.S. banks to the Entente states, and of defending maritime trade with Europe against the submarine offensive besieging France and England. Above all, it was imperative to neutralize the trap into which the Old Continent had fallen with a global conflict that would have led to the hegemony of the Russian autocrats if the Entente had won, or to the domination of an arrogant and warmongering Prussia. A big problem in any case. After the sinking of the Lusitania, the fall of the Tsar in February 1917 was the death knell for the pacifists. On April 2, 1917, the U.S. Congress declared war on Austria and Germany. However, a year had to pass before the Western Front felt the weight and power of the American army. The United States was unprepared and had to arm itself. The Prussians, having relieved the Russian front, tried to take advantage of the delay to deal the final blow to France. They came close to victory, but in the end, they gave up completely and were forced to ask for an armistice.

At beautiful Versailles in June 1919, France and England, embittered by past fears and as greedy as ever, feasted on the carcasses of the defeated, imposed exorbitant war reparations, divided up territories, created new nations and restored old ones, displeasing the allies and frightening the Americans, who closed themselves off. Wilson could not even get Congress to approve membership in the League of Nations, which he had envisioned, because of fierce Republican opposition. This rejection was the death knell for the utopian idea of preventing wars through reason and good will.

Goddard had hoped to sell his skills to the U.S. Army, offering the military the mirage of a new and powerful weapon. His project was considered so sensitive that it required special security measures. Thus, the frail Robert was sent to the hermitage of the Mount Wilson Observatory in California: a place that would soon become the cradle of modern cosmology, thanks to the discoveries of Edwin Hubble. But due to the relative brevity of the American war effort and a relapse of the tuberculosis that had struck him five years earlier, he was unable to accomplish anything significant, except for the project of a portable rocket launcher that anticipated the concept of the bazooka. The idea was developed by a collaborator in time to be used in the Second World War, on the European front and in the Pacific.

Instead, the financial support came from the Smithsonian, a federally funded institution for education and scientific research, to which Robert had approached with a project to make groundbreaking observations of the upper atmosphere. He was explicit in his proposal. By means of a kind of multi-stage rocket, loaded with solid propellants to be ignited in succession, he intended to reach an altitude of 370 km, also carrying some scientific instruments to make measurements *in situ.*

The Smithsonian financed the research with a total of $5000 to be distributed over five years: a sum whose purchasing power is equivalent to 100,000 dollars today. Clark University also contributed a substantial sum, $3500, plus a few rooms at the edge of the campus where the experiments could be carried out. That was in 1917. Three years later, the investment produced a fundamental treatise on *A Method of Reaching Extreme Altitudes*, a work of astronautics on a par with the study published 14 years earlier by Tsiolkovsky, but this time written in a language understandable to many. Against the will of the author and the financiers, it didn't even contain the latest results on solid fuels, which, to be honest, were rather disappointing.

The response was significant, both in the academic community and among the general public, mainly because Goddard, usually cautious and measured in his statements, had ventured to calculate the minimum "*mass [of solid propellant] required to lift a mass of one pound to any desired altitude*". The purpose of the calculation was to determine the conditions for a rocket to escape the Earth's gravitational field and then fall to the Moon. What should have remained an innocent academic exercise became a ticking time bomb because the author had added a fascinating supplement to the discussion. Eager to provide the reader with a method that would show the actual impact of the projectile on the surface of the satellite, should the experiment ever be carried out, he imagined using the explosion of a cartridge placed at the nose of the rocket, and for this purpose estimated how much gunpowder would be needed to see its flame against the dark background of the New Moon through a large terrestrial telescope.

He was immediately dubbed the "*Moon Man*" and ridiculed in the press. The most venomous attack came from the *New York Times*, which on January 13, 1920, in an editorial titled *A Severe Strain on Credulity*, accused him of not knowing "*the relationship of action to reaction, and of the need to have something better than a vacuum against which to react*", even though he was a professor of physics at the university and in the good graces of the prestigious Smithsonian Institute. "*Of course*", the anonymous reporter concluded, "*he *only seems to lacks the knowledge ladled out daily in high schools"*. Even Einstein, by now very popular in America, was brought into the discussion, arguing that only he and a small handful of other minds were licensed *"to deny a fundamental law of dynamics"*. It may seem like a compliment, but it was also an ironic jab at the great Jewish scientist who had dared to question the sanctity of Newton.

Goddard took it very badly. He tried to respond by writing some popular science articles, in an attempt to regain the prestige tarnished by the criticism. Then he gave up and became even more reclusive. The consequences were significant and to some extent dramatic. Thanks to the clamor around his name, some Germans, who, as we shall see, were beginning to think of rockets as a way out of the disarmament imposed on them at Versailles, discovered a rich source of inspiration.

Had it not been for the many aids provided by Goddard until he realized the dangers inherent in Nazism, the V2 rockets might never have flown over the skies of London to sow death. Nor would there have been the opportunity, on both the American and Soviet sides, to harvest knowledge from the Fuhrer's scientists. And who knows, there might not even have been a race to be the first to touch the Moon for lack of competitors, and thus not even the *New York Times* apology to Goddard published the day after the Apollo 11 launch.

Bitter and angered, Goddard threw himself into the arms of the military for the next four years. He took refuge in a Maryland munitions factory to devote himself to designing anti-tank and anti-submarine missiles. Then, back at Clark University, he returned to his old passion. As early as 1922, he had decided to replace solid propellants with liquid ones. In March 1926, as we now know, he carried out the first flight with relative success. The passive control of the rocket in the manner of Congreve did not promise anything good. So, just a month after the historic launch, he began experimenting with active controls based on movable pallets to divert the flow out of the nozzle. A kind of rudder controlled by an inflexible helmsman, a gyroscope.

And 1929 came. The only launch of that year had a meager result. The rocket rose only 27 m before falling back to the ground. Further launches were impossible due to lack of resources. The news ended up in the *New York Times*, testimony to a vigilant journalism and a society interested in scientific and technological advances. Fortunately, the article was read by Charles Lindbergh. The intrepid aviator, who two years earlier had crossed the Atlantic reaching Paris in one leap, was looking for new ideas about flying. Perhaps the Worcester professor could help him? After some positive character checks, Charles decided to call Robert. They met and liked each other right away.

We don't know what they said to each other. Probably their conversation can be faithfully summarized in these lines: "*What can I do for you, Bob?*". "Well, Charles, *what I need are sufficient funds to move forward*". "*I have some powerful friends. I'll see what I can get and I'll let you know*".

In fact, Lindbergh's fame would have been a great skeleton key to open the hearts of politicians and financiers if it weren't for the simultaneous Great Depression that began with Black Thursday on Wall Street, October 4, 1929. There was no more money and the few who had it held on to it. The economy stagnated while the poor, getting poorer, literally died of want. Finally, the following year, the aviator-hero found support in Daniel Guggenheim. He was one of the sons of the patriarch Meyer, who had come from German-speaking Switzerland very poor and had made a fortune in the New World in the metal mining market. The far-sighted financier's generous offer, covered by the *Fund for the promotion of aeronautics* that he himself had established, amounted to $100,000 over four years, the equivalent of $1.5 million today. It was the beginning of a long relationship between Goddard and this family of patrons.

With this substantial nest egg in his pocket, Robert thought of moving to a more suitable place for his experiments, sufficiently deserted and with a good climate to conduct the launches safely while staying away from prying eyes. Leaving the cold of Massachusetts might also benefit his health, which was always in a delicate balance. He chose a ranch near the town of Roswell, in the former Mescalero Apache territory of New Mexico. Two decades later, General Leslie Groves, seeking a top-secret home for the team of physics geniuses assembled to build and test "*the bomb*" to break the Japanese resistance, would make a similar choice.

Two peaceful years passed. Goddard worked intensely, kept sporadic contact with his German colleagues Oberth and von Braun, and opened himself as little as possible to them. Certainly a visionary, but not a worka-holic fanatic, he did not neglect his family and friends, outdoor barbecues, walks, long bridge games, and movies. There were three good starts in 1931 and only one, a partial failure, the following year. Then things got worse.

Struck to the heart by the kidnapping and murder of his young son, Lind-bergh left the scene. To lick the cruel wound, he and his wife sought some peace in Europe. They stayed there for a lustrum, getting too close to the new, glittering Germany of Adolf Hitler, according to American public opinion. In the U.S., Daniel's death and the Depression dried up the Guggenheims' spigot, forcing Goddard, broke and without a patron, to return to Worcester. There he survived on a meager professor's salary from Clark University and a small grant from the Smithsonian. But the critical situation lasted only two years, and in 1933, thanks to the renewed generosity of the Guggenheim, he was able to return to Roswell, where he resumed the interrupted series of launches. The creative vein of this lone genius was not yet exhausted. Goddard worked to refine the details and increase the power of his engines. In particular, he was looking for methods to simplify and make more reliable the way the combustion chamber was fed. What was not already commercially available had to be designed, built, and tested. In 1936, he decided to publish a summary of his work in an article entitled *Liquid-Propellant Rocket Development*, another milestone in astronautics.

By now, his creations, four meters tall or more, had the shape of modern rockets, with a tapered body and four tail fins. Beautiful and functional, with their various equipment, they anticipated many of the solutions that were later adopted, *mutatis mutandis* (made the appropriate changes), in the large rockets destined to realize the dream of flying high in the sky to the Moon. But in none of the thirty-five launches that make up this pioneer's palmares did the rocket powered by gasoline and oxygen go higher than a modest handful of kilometers.

"*Moon rocket misses target by 238,799 1/2 miles*", read a Worcester newspaper's headline on the failed 1929 launch. A remark as ferociously ironic as it was unfair. Although he had not even touched space, Goddard had set many of the conditions for someone else to do so after him.

He died of throat cancer in Baltimore on August 10, 1945, at the age of sixty-two. Before fate cut his thread, he had time to see the documentaries of the devastation Oberth and von Braun's V2s wreaked on European cities. The nightmare that someone might misuse his work had come true. Five years earlier, while America was still watching the umpteenth suicide of Europe from its windows, without taking an active part except to enrich itself, Goddard, together with Henry Guggenheim, had tried to sensitize the high spheres of the U.S. Army and Navy to the danger that the Nazis could build supersonic flying bombs, perhaps capable of crossing the Atlantic. The Pentagon's top brass didn't believe him. In fact, they probably chuckled, thinking that Goddard was confusing the virtues of his fleas with those of a pole vaulter. Eventually, he was given a position as director of research in the Office of Aeronautics, with the task of designing propellers for so-called jet-assisted takeoffs. Once again, he was expected to stay where air helps you fly.

After the war, his book *Rocket Development: Liquid-Fuel Rocket Research, 1929–1941* was published posthumously in 1948. Few took notice. But, when the space race exploded, the figure of Goddard was resurrected from relative oblivion along with his writings. Each of the two competing coun-tries wanted its own champion. Just as the USSR had done with Tsiolkovsky, the U.S. gave its scientist prestigious commemorative honors, even naming after him NASA's new Space Flight Center in Greenbelt, Maryland. In 1960, after a lifetime of begging, Goddard received a million-dollar settlement from the Federal Government for the use of his more than 200 patents in the U.S. space program's military rockets and missiles. Enjoying the large sum, however, were his widow and the Guggenheim Foundation. So goes the world!

The Nibelung Saga

Everyone is a Moon, and has a dark side which he never shows to anybody.
Mark Twain
Research is what I'm doing when I don't know what I'm doing.
Wernher von Braun

On the evening of Tuesday, October 15, 1929, just nine days before the ruinous stock market crash on Wall Street, the *Ufa-Palast am Zoo*, the large cinema owned by Universum Film AG and located a stone's throw from Berlin's *Zoologischer Garten* in the Western part of Charlottenburg, was brightly lit, as it always was for a premiere. Nearly two thousand people, more than the large hall could hold, waited in the cold to be admitted to see the projection of Fritz Lang's latest silent film, *Woman in the Moon.* Dressed in a black coat, long white scarf, and bow tie, the director, now a star of German cinema thanks to such masterpieces as *Dr. Mabuse the Gambler* and *Metropolis*, was there with his wife, the beautiful and sophisticated Thea von Harbou, his muse and the author of the novel from which the film was adapted.

The plot consisted of the usual intertwining of love, loyalty, greed, and cowardice, set against the backdrop of a journey to the Moon to get at the rich gold mines that a visionary scientist had hypothesized might be hidden in the bowels of Earth's satellite. A saccharine adventure far removed from the heights of creativity already touched by Lang. But even before its release, the movie had aroused a certain curiosity for the novelty it contained. In order to tell the story of the voyage to the unusual sidereal Klondike aboard the Friede rocket, the namesake of the film's heroine, without the naiveté and errors that

© The Author(s), under exclusive license to Springer Nature
Switzerland AG 2024
M. Capaccioli, *Red Moon*, Springer Praxis Books,
https://doi.org/10.1007/978-3-031-54760-7_4

had characterized the classic novels of Verne and Wells, the almighty Lang, an avid science fiction fan, had demanded a rigorously scientific approach. To this end, he had sought the technical advice of a German-speaking engineer with a Romanian passport, who had gained fame and respect with two monographs on rockets and their future use. More than the plot of Lang's movie, it is the story of Hermann Julius Oberth that interests us. Indeed, if Wernher von Braun is to be recognized as the father of the enterprise that took Neil Armstrong and Buzz Aldrin to the Moon, Oberth, a man *"tall, thin, straight, with thick dark hair, broad chin, black mustache, and those lively eyes that are the prerogative of an alert mind"*, was certainly its grandfather.

When he was born on June 25, 1894, his hometown, Hermannstadt, was located within the borders of the Austro-Hungarian Empire, in that vast plateau surrounded by the Carpathian Mountains that the Romans called Dacia and that in the Middle Ages took the name Transylvania, meaning "land beyond the woods". A country of lakes, greenery and castles perched on impassable mountain ridges—including that of Count Dracula—that was systematically colonized from the twelfth century onwards by German-speaking peoples, the Saxons of Transylvania. In 1920, however, as a result of the Treaty of Trianon, which redrew Hungary's borders after the collapse of the Habsburg Empire, more than half of the region was given to the Kingdom of Romania. The city of Hermannstadt was renamed Sibiu, and the 26-year-old Hermann became a Romanian citizen. In reality, however, he was and remained primarily German, by family, education and, after the Second World War, by choice.

His father, a wealthy physician-surgeon and director of the local hospital, was in love with his profession and, as is often the case with successful men, hoped that his son would follow in his footsteps. Not wanting to disappoint his father, or, more likely, not having the strength to rebel against him, Hermann enrolled in the Medical Faculty of the University of Munich in 1912. However, his true passion was something else, since, at the age of eleven, he had read during a long convalescence in Italy two novels by Verne, procured from his father, which talked about voyages to the Moon. Three years later, still in high school, he had even designed a model of a rocket for interplanetary travel, coming up with a multi-stage solution using liquid propellant to progressively achieve the necessary speed. By some simple calculations, he had realized that the bullet-ship envisioned by the French writer could not work, because the monstrous impulse exerted by the cannon at the moment of firing would prove fatal to the human crew, who would be crushed like a walnut by the unbearable pressure. For the moment, nothing more than a teenage passion, fueled by reading as much as possible and daydreaming. The kind that often evaporates with growth, with first loves, and with the first harsh confrontations with the reality of life. But the reflections of this young man with an intense and deep gaze showed an extra gear than usual.

The leap from forest town to imperial metropolis, a rich and joyful melting pot of art and culture, was significant. In Munich, the carefree and concrete nature of the Bavarians was mixed with Prussian austerity, creating a pleasant and stimulating mixture in the pervasive and constant smell of good beer, which drives away bad thoughts. The Academy was thus a veritable Athens of the sciences, both theoretical and applied. Imprisoned by the promise he had made to his father, Hermann attended medical lectures, but also those of mathematics and physics at the prestigious *Technische Universität*. If the war hadn't upset the delicate balance, his passion for space travel might have remained at the level of a hobby to be cultivated in the little free time left to a good and conscientious doctor.

At the outbreak of the conflict, he was twenty years old, the right age to wield a musket. He was immediately drafted into the German infantry and sent to the Eastern Front to fight the Russians, where he was wounded. Transferred to the military hospital at Schässburg, not far from home, probably thanks to his father's good offices, he remained there as a medic, realizing firsthand that this was not the kind of life he wanted for himself. In his spare time, he thought about his dream of going to the Moon. There were two main problems to solve. Building a spacecraft capable of making the journey and testing the crew's ability to withstand the accelerations and lack of gravity. He went on to design liquid-fueled rockets, breaking them down into chains of tanks that could be detached as they were emptied to gradually reduce weight. Exactly the multi-stage solution that would later carry *Sputnik*, Gagarin's *Vostok*, and Apollo 11 into the deep sky. He even managed to show his detailed rocket project to the Prussian Minister of War, Hermann von Stein, who, as a former artilleryman, had a fine enough palate to appreciate it. The medical environment then stimulated him to perform experiments on himself to analyze the effects of weightlessness on the human body. Before the armistice, in the summer of 1918, he took a wife, with whom he had four children. Two of them died violently during the Second World War. Tragedies shared by many fathers of a country sent to slaughter by a bloodthirsty madman.

With the peace, he returned to Munich, this time to study exclusively mathematics and physics with great masters. The defeat had humiliated, exhausted, and impoverished the Germans, but it had not affected the formidable educational structure of the Prussians, nor had it damaged the urban infrastructure, as would happen during the Second World War.

Germany was still standing and poised for rebirth, though still gripped by hunger and shaken by an internal conflict between authoritarian nationalism and Marxism. In three years, on the same benches where Oberth now sat listening to the wonderful lessons of Arnold Sommerfeld, sublime geniuses of the level of Werner Heisenberg and Wolfgang Pauli would have taken their place.

Then Hermann moved to Göttingen, in the historic university of Friedrich Gauss where David Hilbert was now teaching, to finally arrive in Heidelberg, a center of democratic thought and also a den of conservatives and Nazis. There he hoped to obtain a doctorate in physics by defending a thesis on space flight. But the degree was denied. The committee found the thesis lacking in concreteness: a futuristic essay instead of the traditional solid physics and mathematics dissertation. It discussed, for example, the strategy for exploring the hidden face of the Moon and refueling spacecraft by storing cryogenic propellant tanks in Earth orbit to be docked and used for interplanetary travel or parked around the celestial body to be visited.

Although the Weimar Republic was in complete ideological chaos, shaken by galloping inflation and the violence of ever-growing disaffected factions, the German academy preserved the rigorous and rigid approach forged in the years of the Empire and symbolized by the *Pickelhaube*, the spike helmet. The motto was "facts and not dreams", even if they were plausible. Oberth knew it. "*Our educational system is like an automobile*", he later wrote, "*which has strong rear lights, brightly illuminating the past. But looking forward things are barely discernible*".

To react to the rejection, he did two things. He returned home, to Romania, where the following year, without changing anything in his work, he finally managed to obtain the diploma from the ancient University of Cluj. At the same time, he published the original dissertation at his own expense—apparently he could afford it—in a hundred-page pamphlet in German entitled *The Rocket into Interplanetary Space*, and rushed to send a copy to Goddard. It was not just a *beau geste* to reciprocate the kindness of the American who, at his request, had reluctantly sent him a copy of the 1920 essay on *A Method of Reaching Extreme Altitudes*. It was also a ploy to firmly stake his claim. In the acknowledgments, after starting with a sort of *excusatio non petita* (unsolicited excuse): "*Goddard's work was received just as this was going to press*", he declared with a certain brazenness: "*My theoretical approach is supplemented by his practical work*". That is, the brawn and the brain. Not bad for a beginner!

Faced with this unfortunate statement, perhaps inspired by Hermann's frustration over his academic failures, Robert, already suspicious by nature, was convinced that *"that German Oberth"* had plundered his work and became even more cautious and closed. A real disaster for the United States, which, with the reluctance of the Worcester genius, missed the opportunity to take off far before anyone else on the rocket front. It was also a disaster for the rest of the world, which did not benefit from the valuable experience that the brilliant American had gained in this field. Science and technology, in fact, behave like multi-stage rockets. Each successive element inherits the thrust of all the previous ones. Partly because of the arrogance of a rising young man, Goddard refused to act as the first stage and closed the communication channels.

On the other hand, Oberth's booklet had the positive effect of awakening interest in space travel in the German-speaking world. With a scientific approach, the author painted a future consisting of permanent stations in Earth orbit. Large rotating rings to simulate gravity by centrifugal force, to be supplied by a constant flow of small shuttles. Authentic celestial ports with a service function for interplanetary spacecraft but also platforms for telecommunications and meteorological monitoring. Enough to arouse the curiosity of young minds in years of great political confusion and serious social upheaval. Unfortunately, the consequences of this proselytism would have become another drama for Germany and the world. Not for nothing is it said that sometimes the best is the enemy of the good!

In fact, Oberth's seed germinated in the *Verein für Raumschiffahrt* (VfR— Society for Space Travel), an association of people interested in rocketry and interplanetary flight founded in June 1927 by 30-year-old Johannes Winkler. He had long been involved in the design of rocket engines for use in space, a hobby that would lead him two years later to a position with Hugo Junkers' aviation company in Dessau, Germany, to develop jet engines to assist in the takeoff of jet aircraft.

Within twelve months, the association had gathered more than five hundred members. A sign of the relevance of the proposed theme, but also of the scarcity of other values in a society still in disarray and in search of a renewed identity. For seven years it also published a magazine, *Die Rakete* (The Rocket), and it was the nest from which Wernher von Braun took flight, one of the two champions of the carousel for the conquest of the Moon. About the other champion, the Russian Sergei Korolev, we will speak later.

Oberth, who had inspired it, played the role of noble father and advisor to VfR, mostly from afar. Rejected by the German academy, he had rejected Germany for the time being and—he who dreamed of heaven—had settled for teaching mathematics and physics at a high school in Mediaș, a spa town about fifty kilometers from Sibiu. Whenever he could, however, he visited old friends in Munich and new ones from VfR Berlin.

Sometimes he would agree to be consulted about the world of rockets, as in 1928 when, together with another member of the VfR, the popularizer of science Willy Ley (1906–1969), stayed in Potsdam, in the Babelsberg district, in the studios of Universum Film AG, to assist Fritz Lang during the shooting of *Woman in the Moon*. It was a way to scrape together some money and, above all, a golden opportunity to popularize his vision of the future.

For a while, Oberth even hoped to promote the film by using the funds provided by Lang to build a real rocket that could be launched up to 10 km high over the Baltic Sea. He had only four months to build a two-meter-high model fueled by gasoline and liquid oxygen, similar to the one flown by Goddard in 1926. Technological and managerial inexperience quickly doomed the project. Hermann then opted for a confused hybrid solution of a giant powered by experimental fuel. The program died before it was born. A second flop that convinced him to return home, but not before ensuring that the launch of the rocket model used in the movie was sufficiently convincing.

He reappeared in Berlin to attend the premiere. The feature film was a success and served to draw further attention to rockets and space. Its realism, so different from the mocking parody of Verne's novel that another great, Georges Méliès, had packaged in the 1902 short *Le Voyage dans la Lune* (A trip to the Moon), could not go unnoticed. Many still doubted, but some began to truly believe that the new frontier of mankind was right up there, beyond the clouds, where the air ends and empty space begins. Even the first V2 successfully launched from Peenemünde would have a reference to the popular movie imprinted on its base, demonstrating how deep the mark it had left was.

In the same year, 1929, Oberth published an extensive essay entitled *Wege zur Raumschiffahrt* (*Ways to Space Travel*) with the prestigious R. Oldenbourg publishing house in Munich. It was nothing more than a completely revised and significantly enriched re-edition of the dissertation that had appeared six years earlier. It contained formulas, tables, graphs, and projects that finally made quantitative a discipline born on the wave of a dream. Despite the computational difficulties that are now overcome by the use of powerful digital computers, Hermann had managed to get to the heart of the problems and to identify the critical parameters of the various situations.

A way of being and doing happily summarized in a memory dictated by Wernher von Braun, who knew him very well:

Hermann Oberth was the first, who when thinking about the possibility of space-ships grabbed a slide rule and presented mathematically analyzed concepts and designs [...] I, myself, owe to him not only the guiding-star of my life but also my first contact with the theoretical and practical aspects of rocketry and space travel. A place of honor should be reserved in the history of science and technology for his ground-breaking contributions in the field of astronautics.

The knots to untie concerned four aspects of the problem of human space flight not yet verified and therefore the subject of bitter controversies. Is thrust in the vacuum possible?[1] If so, how to reach speeds that can escape the gravitational pull of the Earth? Will astronauts be able to withstand the strong accelerations and adverse conditions of interplanetary space? What is the point of all this? Oberth had an answer for each of these questions. Mathematically and physically founded for the first two, more optimistic than scientific for the third, and simply speculative for the fourth.

His boundless imagination led him to envision a future in which fleets of small "ferryboats" swarmed around truly floating citadels in orbit. They would house the new pioneers and allow them to conduct scientific experiments and monitor the planet. Outposts of a new East India Company for commercial and mining activities extending into the Solar System or fortresses for military control of old Earth. All of this, he wrote, would stimulate the development of new technologies in many different fields that would benefit the quality of life for all of humanity.

This argument, the latter, is still used today to justify the astronomical costs of space exploration, but it continues to divide public opinion. Why expend vast resources to go up there, to the middle of nowhere, when there is still so much to do down here to free people from hunger, misery, and disease? In 1970, after the successes of the Apollo lunar program, Sister Mary Jucunda, a nun surrounded by starving children in her poor mission in Kabwe, Zambia, asked this question again to Dr. Ernst Stuhlinger (1913–2008), the powerful associate director of science at NASA's Marshall Space Flight Center and responsible for expanding the Apollo project to include human exploration of Mars.

The former Nazi scientist, who had already been a faithful servant of the Führer in Peenemünde and was therefore presumably equipped with sufficient stomach for it, did not want to dismiss the provocation and responded with a long open letter in which he dissected the problem. His main argument remained, in an extreme synthesis, an evaluation of the cost–benefit ratio. An investment is successful if the future return is greater than the present commitment. And to make his point, he used an anecdote.

---

[1] The problem had already been addressed and solved by Goddard in his Worcester laboratory.

In seventeenth-century Germany, a kind-hearted Count helped his poor people to survive and to cure themselves of disease. One day he met a strange individual who seemed to be playing with glass lentils. He became curious and decided to finance his activity. "*The townspeople, however, became angry when they realized that the Count was wasting his money, as they thought, on a stunt without purpose*". It was instead an investigation into the causes of diseases with a microscope. "*By retaining some of his spending money for research and discovery*", concluded Stuhlinger, the Count "*contributed far more to the relief of human suffering than he could have contributed by giving all he could possibly spare to his plague-ridden community*". No one knows how Sister Mary Jucunda reacted to this answer, wise and cruel. Perhaps she read it while lulling one of her black children with big eyes hollowed out by hunger and fever.

The VfR was an amateur society not different in structure, organization, and purpose from similar associations that had flourished especially in Russia and the United States since the twenties. But it was the one that had the most important consequences for the history of rocketry. Its members dealt with everything, from pure entertainment to real studies and rocket projects. In 1926, for example, Oberth and the South Tyrolean Max Valier, an imaginative and talented writer with scientific training and interests, practiced redesigning Verne's Moon gun to make it functional, if not feasible, and essentially usable. Two years later, the same Valier—who would soon die in a rocket explosion, the first victim of the long journey to the Moon—collaborated with Fritz von Opel in the realization of the experimental car, the Opel Rak.2, which broke the speed record in Berlin on May 23, 1928, reaching 230 km/h thanks to the thrust of 24 rockets. It was driven by the same Opel who wanted to promote his automobile company with this risky sporting venture.

The mixture of dream and business displeased Valier and especially the purists of the VfR and contributed to sowing discord within the association. It was necessary to give the activities a new direction, to make them more professional and concrete, building and testing rockets. Resources and a headquarters were needed.

The latter were identified in Reinickendorf, on the northern outskirts of Berlin, in an area of 122 hectares that had already been used as an ammunition dump during the First World War. Between old concrete buildings, now ruined, an open space was carved out, marked by a pompous sign that read, in gothic letters, *Raketenflugplatz Berlin* (Berlin Rocket Airfield). Here, with the approval of the city hall, the first experiments began. The failures were numerous. But the few successes made it possible to reach a record altitude of 1000 m in 1932. In this poor environment, amidst debris and crumbling buildings, the dream of reaching the Moon really took off, only to become a reality within forty years.

The winds of change were blowing. Germany was gradually re-emerging as an economic and political power. In 1926, it was admitted to the League of Nations thanks to a reconciliation with France that would earn the two negotiators, Chancellor Gustav Stresemann and Prime Minister Aristide Briand, the Nobel Peace Prize. Three years later, a plan led by an American industrialist, founder of the RCA record company, had initiated a drastic reduction of the war reparations imposed on the Germans. A pragmatic move that recognized the exorbitance of the punitive measure demanded by the French, still in shock after seeing the Prussians knock on the doors of Paris twice in less than half a century, and also sought to exorcise an act of rebellion in a starving nation and thus save at least part of the credits.

Even the rising tide of Nazi partisans seemed to be under control. In the 1930 elections, the followers of the "Bohemian Corporal"—as Adolf Hitler was scornfully referred to by the aristocratic President Paul von Hindenburg, a true relic of the defeated Reich—had won 18% of the seats. But with a mere 26%, the *Sozialdemokratische Partei Deutschlands* (SPD—Social-Democratic Party of Germany) remained the largest force in the Reichstag. A fragile barrier that would not last long. Meanwhile, America, guarantor of victory, had further retreated into its own shell to lick the wounds left by Black Thursday on Wall Street. A rampant leprosy of the economic system ready to offer an unexpected assist to the Nazis.

Taking advantage of the diminished attention of the U.S., which was preoccupied with the problems of the recent Wall Street meltdown—a fact that would soon give the Nazis an unhoped-for advantage—the Germans cautiously sought to rearm, circumventing and in some cases violating the prohibitions imposed on them by the Treaty of Versailles. The clauses contained in Section III of the text drafted by the victors: "*The armed forces of Germany must not include any military or naval air force*", had literally amputated military aviation, humiliating the now deceased aces like the Red Baron or the surviving and angry ones like Hermann Göring.

Even solid-fuel rockets were banned because they were associated with artillery shells like Krupp's Big Bertha or Kaiser Wilhelm's Gun, which had even managed to bomb Paris. But nothing was written about liquid-fueled rockets because they did not yet exist. An understandable oversight that became a delicious loophole for legal "technological" rearmament and made the activities of the *Verein für Raumschiffahrt* quite attractive. At least that's what its leaders hoped, perennially hunting for sponsors.

In the early months of 1931, VfR president Johannes Winkler, armed with his experience at Junkers and with the help of private sponsors, built his own liquid-fuel engine. His rocket attempted to fly twice, crashing miserably to the ground each time. However, it was a first in Europe. The unilateral initiative displeased the other members of the association, who felt cut off from an enterprise that was part of the VfR's mission. Almost an insult from the person who had founded and led the club. There was some turbulence, physiological in associations of enthusiastic and visionary volunteers. Then optimism prevailed. If it could be done—everyone agreed—it had to be done well, and all together. The most active members were Rudolf Nebel, the oldest of the group, a pilot in the Luftwaffe during the First World War, an enthusiastic empiricist and a skilled fundraiser; Klaus Riedel, a mechanical engineer who worked for an automobile equipment company; and a young man not yet twenty, handsome, polite, and taciturn. He was said to be a gentleman with powerful connections. His name was Wernher von Braun, and he had already worked as an apprentice under Oberth on an innovative engine for the rocket (called *Kegeldüse*, cone nozzle, because of its conical combustion chamber) whose launch was to accompany the premiere of Lang's film. An idea in the manner of Oberth, grandiose, formally correct, but impracticable.

The three from VfR had decided to reduce their demands and to concentrate on a minimum rocket, the MiRAK, which had the potential to really fly. With the total lack of money, they could not have done more, given the numerous knots to untie. The first and most serious problem was the cooling of the combustion chamber, whose resistance was severely tested by the enormous heat. Goddard had solved this problem brilliantly by directly using the cooling effect of liquid oxygen. Inspired by car engines, von Braun and his colleagues opted instead for a water circuit.

After all kinds of tests and resolving some misunderstandings within the group, MiRAK III flew on March 14, 1931. A sort of arrow, 3.5 m long and 10 cm wide, with a full load weight of 20 kilos, half of which was fuel. To prevent the heat of the exhaust gases from exploding the fuel tanks, they had been moved to the rear, as in Goddard's original project. First the liquid oxygen and then the alcohol.

Willy Ley, who served as the group's spokesperson and later, after escaping Nazi Germany in 1935, became a famous science writer in America, described the event this way:

> *The rocket took off with a wild roar, hit the roof of the building, and raced upward at an angle of about 70 degrees. After about 2 seconds, it started to loop, climbed some more, dumped all the water out of the cooling jacket, and came down in a power dive. While it was diving, the wall of the combustion chamber – no longer*

cooled – gave way at one point, and with two jets spinning it around, the thing went completely crazy. It did not crash because it ran out of fuel just as it was coming out of a power dive near the ground. In fact, it almost made a landing.

A partial fiasco. Its successor, the Repulsor II, fared a little better. It did not rise any higher than its predecessor, but it traveled over half a kilometer horizontally and was recovered intact thanks to a parachute. The expedient allowed it to be reused for dozens of launches, which created a lot of excitement in the VfR. It was time to let the world know. Ley took care of it again with an article in the press and by communicating with other rocket fans. The intention was to attract a wealthy backer. Instead, it attracted the attention of the military, and so Captain Walter Dornberger (1895–1980) came on the scene.

As early as 1930, this artillery officer had been assigned by the *Reichswehr* Armaments Office[2] to a secret project to develop liquid-fueled rockets capable of competing with conventional weapons. Dornberger was the right man in the right place at the right time. A lieutenant in World War I, he had been captured and spent two long years in a French prison camp, almost all of it in isolation because of his repeated attempts to escape. When he finally returned to his homeland, he completed his university studies, graduating in 1930 with a degree in mechanical engineering from the *Technische Hochschule* in Charlottenburg (followed by an honorary doctorate in 1935). A brave and stubborn soldier and a skilled technician.

For his new assignment, he had settled in Kummersdorf, an isolated area about 25 km south of the capital, where he began to study the problem. Determined not to reinvent the wheel unnecessarily, he went to Reinickendorf in the spring of 1932, together with his commander, Captain Ritter von Horstig, and the head of the Ballistics and Munitions Office, Colonel Karl Becker, to watch a launch by the "boys" of the VfR. Three experts to test the enthusiastic statements of Ley. At that time, Dornberger later recalled, the world of rockets "*was a sphere of activity beset with humbugs, charlatans, and scientific cranks*" and hardly populated by really capable people.

The demonstration was a fiasco. Despite the discouraging results of the inspection, Dornberger offered the VfR a small contract for another demonstration launch, which took place at the Kummelsdorf military range and went much better than the first. The rocket flew 1 km high and three times horizontally. However, Becker had no intention of spending the scarce public funds to entertain the members of the VfR. After some back and forth, his

---

[2] Defense of the Reich: name given to the German armed forces from 1919 to 1935, when it was changed to *Wehrmacht*, Defense Forces.

proposal took the form of a close collaboration, aimed at and bound by military secrecy. Dornberger later wrote in his memoirs that they intended to put an end once and for all to theories, unsubstantiated claims, and vain fantasies, and to come to conclusions based on sound science.

Newcomer Wernher wanted to accept the contract and even signed it personally. But the majority of the VfR's elders refused. Among other things—they argued—as the rockets reached higher and higher altitudes, it became risky to operate from a base so close to the city. The municipality of Berlin could have terminated the lease at any time, leaving VfR without a roof over its head. The ensuing discussions contributed to the death of the association, which was finally closed in early 1934 with the approval of Propaganda Minister, Joseph Goebbels.

For one year, the music in Germany had radically changed. The *Nationalsozialistische deutsche Arbeitpartei* (NSDAP—National Socialist Party) had won the elections of 1932, albeit by a small margin, and in January 1933 the old Hindenburg, now with one foot in the grave, had to give way and hand over the chancellery to the "*Bohemian Corporal*", who quickly managed to gain full powers. A year later, on August 2, 1934, after clever maneuvering and cynical violence against friends and enemies, Hitler assumed the position of Führer and Chancellor. The Third Reich was born, a totalitarian regime animated by intentions of revenge and supremacy *über Alles*. All the best Aryan minds had to work together to achieve the goal: even the rising star von Braun.

He had now made his choice and moved to Kummersdorf, leaving the university classrooms where he was completing his doctoral thesis. He was welcomed with open arms, for he had already shown himself to be a champion. The officers with whom he had interacted had the intelligence, the preparation, and the open-mindedness to fully understand it. He also belonged to a very prominent family in Berlin, which certainly played into his hands.

Wernher Magnus Maximilian von Braun was born on March 23, 1912 in Wirsitz, in eastern Germany. In 1920, when the city became part of the territories returned to Polish sovereignty under the Treaty of Versailles, most German-speaking families were forced to emigrate to Prussia. Among them were the aristocratic von Brauns, who moved to Berlin, where Wernher, the second of three brothers, grew up and studied.

The father, Magnus, was a remarkable man in his own right. A Junker with ancient roots, he had served as a spokesman for Kaiser Wilhelm II. After the war, he held the same position in the SPD. In 1932 he was appointed Weimar Germany's Minister of Food and Agriculture. But soon after, disliking the new chancellor,[3] he withdrew from the political scene. Wernher's mother, Emmy Melitta von Quistorp, was an elegant woman, refined and cultured, with direct connections to a number of European royal families. An authoritarian aristocrat and a lover of music and astronomy, she had given her son the novels of Verne and Wells, as well as a telescope to look at the Moon. But what the boy liked best was speed.

"*When I was 12 years of age*", von Braun will recall in 1963, "*I had become fascinated by the incredible speed records established by Max Valier and Fritz von Opel. So I tried my first practical rocket experiment*". He attached a handful of firecrackers to a toy cart and set them on fire. It resembled "*the attempt made by a Chinese named Wan Hoo in 1500*". The object flew away at a high speed, causing panic and a minor disaster, so much so that he was taken into custody by the local police. He was released only upon the arrival of his father, who had rushed to the gendarmerie. A youthful prank that ended well, or perhaps just a cleverly crafted anecdote to humanize the first life of a character who "lived twice". The first part lasted until 1945 and was stained with the blood of many thousands of innocents.

Despite his many interests and a very Prussian attitude toward his student duties, Wernher did not do well in school. Especially in scientific subjects. He loved music (he was a piano student of the composer Paul Hindemith for a while) and dreamed of adventure. Then he was struck by his encounter with Oberth's book, *Die Rakete zu den Planetenräumen* (The Rocket to the Planetary Spaces), fascinating but full of incomprehensible formulas, and fell off his horse on the road to Damascus. He had understood that his karma commanded him to reach the Moon and that this required a careful and deep study of mathematics and physics.

Said and done, he became a model student. In 1930 he entered the *Technische Hochshule* in Charlottenburg. He was an intern in a locomotive factory when fate wanted him to meet his idol. Oberth kept this enthusiastic and helpful boy with him for a while. Then he directed him to the VfR, where Wernher received the first practical rudiments of rocket technology and the invitation to work for Dornberger in Kummersdorf. When he first arrived at the military space center as an army technician, but without a uniform, his disappointment was great. The facilities were even more miserable and bare than the amateur laboratories of the VfR at the rocket airfield. But the die had been cast. Wernher rolled up his sleeves and quickly built ever more powerful and efficient rocket engines, which were initially tested only on the bench. Taking risks like amateurs with nothing to lose would have been suicide in

---

[3] "*If Hitler takes power*", he had presciently said, "*it will be the end of Germany*".

a win-at-all-costs environment. He feared not for himself, but for his dream. To save it, he was willing to do anything, even sell his soul to the devil, as he would later do. He reinvented many of the solutions already found by Goddard, including cooling the combustion chamber with liquid oxygen, and gave them the German brand of perfection that was missing from the lone American inventor's achievements.

Finally, the time had come for flight. After the first test failed by a hair, von Braun was able to launch two Aggregat 2 rockets from the island of Borkum in the North Sea. The area had been chosen to provide a sort of firing range wide enough to minimize the dangers to people and human activities. It was a success, but was followed by a serious accident in which three people lost their lives. Experiments also began to install liquid-fueled rockets in aircraft. A project strongly supported by Ernst Heinkel, a powerful aircraft manufacturer and influential member of the Nazi Party.

Although the subsequent launches of the Aggregat did not go so well, Wernher was at the height of his happiness. He had at last received his doctorate with a thesis that had been declared a military secret, he was leading a top-notch research group, he had called in his best old friends from the VfR days, and he was sure that he would be able to learn from every mistake and achieve success. The army, however, had no intention of waiting. They demanded a working weapon and they wanted it now. For his part, Minister of Aviation Hermann Göring, always on the hunt for personal affirmation, was eager to get his hands on the toy.

In 1938, also at the urging of Wernher's influential mother, the group received generous funding and moved to Peenemünde, on the far north side of the Baltic Sea island of Usedom, near the mouth of the Peene River: "*It's the perfect place for you and your friends; your grandfather used to go there to hunt ducks!*". She knew the area well, having been born in Anklam, a town in Western Pomerania. The base, which was quickly equipped with barracks, technical facilities, launch pads, and large workshops for the production of rockets, was codenamed *Heimat Artillerie Park 11 Peenemünde* (Home Artillery Park 11). It was a strategic choice that provided a large and secure testing ground and greater protection from prying eyes.

International espionage now targeted the activities of the Germans, whose massive rearmament was already feared. The British, French, and Russians hoped to steal the secrets of the new wonders Hitler was packaging for war. Rocket technology seemed to have a rosy future in an impending conflict, and the Führer, though without particular enthusiasm despite his passion for all kinds of new weaponry, demanded that it be formally incorporated into the activities of Germany's new course. As director of the army's rocket center, von Braun was urged to enter the Nazi Party. *"My refusal to join the party would have meant that I would have to abandon the work of my life. Therefore, I decided to join."*, he would later explain, adding that his membership did not involve any political activity on his part.

Shortly thereafter, on September 1, 1939, Hitler, emboldened by the inaction of the French and British governments, launched the invasion of Poland and war broke out again. It was a disaster for the world and for von Braun, who saw his funds and, most importantly, his personnel cut and diverted to the front. Thanks to the powerful friendships of Dornberger, now a major general, the situation was quickly reversed. Peenemünde even received an allocation of 3500 Wehrmacht soldiers. Wernher was thrilled. More brains were needed to create the desired flying bomb. So he assembled a team of specialists, the "Peenemünders", from the country's various university institutes. Whenever he had a problem, he went to see them one by one in a small plane that he piloted himself.[4] A fury, in the whirlwind of a *blitzkrieg* that seemed already won.

Only England, protected by the Channel, resisted. To overcome this last obstacle to the Third Reich's total victory, it was necessary to destroy the Royal Air Force. Göring's Luftwaffe was supposed to do it, but the heroism of the British pilots and the mistakes of the arrogant hierarch extinguished the Führer's ambitions to quickly end the war on the Western Front. Hitler's plan was to have free hands to denounce the Molotov-Ribbentrop pact of non-aggression between Germany and the USSR and pour a torrent of fire on the Bolsheviks, whom he hated in his delusions as much as the Jews. The momentum of the first months was gone.

The slowdown in the military operations was a breath of fresh air for von Braun, as it gave him more time to prepare his flying bombs. To stay in the game, however, he had to agree to become an officer in the *Schutzstaffel* (SS), the Nazi Party's paramilitary organization, and don the black hat with the skull symbol, the sinister *Totenkopf*. It happened in the spring of 1940; a colonel walked into his office with the message that *Reichsführer* SS Heinrich Himmler had asked him to join the SS. Wernher hesitated for a while, then called his military superior, Major General Dornberger, for advice. Dornberger told him that if he wanted to continue his work, he had no choice but to join. *"After receiving two letters of exhortation [from the SS colonel], I finally wrote him my consent"*. He was appointed second lieutenant and eventually promoted to major (*Sturmbannführer*). The von Braun of Peenemünde is often portrayed in the clips of knowledge in pills as a radiant Nibelung in

---

[4] He obtained a pilot's license in 1933.

the service of the Devil, a cold and ruthless genius with no other concerns than the creation of amazing machines. The picture is only partially true, however, because those who serve Satan must frequent Hell, where life is grim and uncertain for everyone, even for the "fallen angels".

A man of dual personality, then, as dual was his judgment (in retrospect) of his Führer:

I met Hitler four times. When I saw him from a distance for the first time in 1934, he appeared to me as a fairly shabby fellow. Later when I met him in a smaller circle [in 1939 and 1941], I began to see the format of the man: his astounding intellectual capabilities, the actually hypnotic influence of his personality on his surroundings. It moved one somehow [...] My impression of him was, here is a new Napoleon, a new colossus, who has brought the world out of its equilibrium [...] In my last meeting with him [July 1943], Hitler suddenly appeared to me as an irreligious man, a man who did not have the feeling that of being responsible to a higher power, someone for whom there was no God [...] He was completely unscrupulous.

Finally, in the fall of 1943, the long-awaited Aggregat 4 (A4) was ready for flight testing. A modern missile, similar to a bullet with four fins, 14 m high and weighing 13.5 tons, capable of carrying an explosive charge of one ton at the tip. After two failed launches, the third one took off in a cloud of smoke on October 3, 1942. When its thrust exceeded its weight, it took off from the launch pad and crashed into the sea at a distance of 200 km after a perfect flight in which it had surpassed an altitude of 80 km. In the latest versions, it would have reached an altitude of 360 km and a top speed of 5200 km/h, making it undetectable by anti-aircraft systems and fighters. This is how it worked. The jet engine pushed it to the preset altitude, into the stratosphere, then let it be governed by the Earth's gravity so that the subsequent ballistic trajectory led it to intercept the target, a supersonic and silent angel of death.

Oberth, the noble father, also drank to the great technological triumph, called to Peenemünde by the student who had become the master. A star had been born, and "*a new era in transportation, that of space travel*", had finally opened; but in the worst way, under the sign of devastation. Dornberger and von Braun celebrated, seemingly overjoyed at their victory, although both knew that the road was still going uphill. Their creature was not yet ready for mass production, as the military, dazzled by this first success and the need for new weapons, would surely demand. More time was needed to solve the thousand open and unknown problems, while keeping an eye on the competition. The Luftwaffe, in fact, was also working on a flying bomb.

Commander of the mythical Richthofen's Circus[5] during the First World War, Hermann Göring did not accept to share the "Kingdom of Heaven" with anyone and he had assigned a brilliant designer to develop the program of the new weapon, taking him away from the study of the Messerschmitt 262, the jet plane that could have given to the Germans the superiority in the skies. Thus was born the *Vergeltungswaffe* or V1 ("retaliation weapon", a name coined by the Mephistophelian Minister of Propaganda, Joseph Goebbels). A hybrid between a rocket and an unmanned airplane, the V1 threw a spanner in the works of von Braun. Among other things, there was a rumor that Hitler no longer believed in the V2 project.

Meanwhile, the tide of the war had changed. Operation Barbarossa to invade the Soviet Union, which Hitler had risked in the summer of 1941 to anticipate Stalin's moves, was turning into a disaster after an initial triumphant advance by the Wehrmacht. Backed into a corner, the Führer began to hope for the miracle of the *Wunderwaffen*. So he agreed to meet with von Braun and Dornberger at his Wolf's Lair, the military headquarters in East Prussia from which he directed his armies on the Eastern Front. He was shown a movie of an A4 launch. The dictator was so enchanted that he changed his mind and ordered the weapon to be produced and deployed as soon as possible. In 1963, von Braun would recall it with these words:

> *I would like to correct an error which you find in a number of stories about the V2, namely that the V2 was Hitler's devilish idea, designed to conquer the world. Up to 1943 Hitler had absolutely nothing to do with the rocket program [...] He simply was not interested. We could not understand it, because he was very much interested in the technical details of all other weapon [...] Not until July 1943, when we finally convinced him with facts, did he see any usefulness in our rockets, and then not as a weapon but as a war-preventive means. "Why didn't I believe in the success of your work?" Hitler asked me. "If we had had this weapon in 1939, we never would have had this war. Now and in the future, Europe and the world are too small for a war. With such weapon available war will become unbearable for the human race". Some hours later, he told me, "I have to apologize only to two people in my life. One is Field Marshal von Brauchitsch. I did not listen to him when he pointed out over and over again the importance of your development. And the second is you. I did not believe in any success for your work".*

In Peenemünde, everyone got to work. As if by magic, the endless difficulties disappeared, leaving room for the most effective cooperation. Workshops were set up for the production of the various parts of the rocket and the

---

[5] Fighter squadron of the *Deutsche Luftstreitkräfte* (German Air Combat Forces) during World War I, led by *Rittmeister* (Captain) Manfred von Richthofen, *der Rote Baron* (the Red Baron).

training of the future "artificers" began. Göring's maneuvers to oust von Braun and leave the field open to his Luftwaffe were in vain. A4 and V1 continued their journey in parallel.

The frenetic activity of the laboratories on the Baltic island had aroused the curiosity of the Allies, alarmed by some photographs and the analysis of a rocket wreckage recovered by the Polish resistance and delivered to the British. A drastic cure was chosen, the same one used for other potentially dangerous Nazi scientific installations: a carpet bombing that took place on the night of August 17, 1943. After outmaneuvering German fighter planes with a feint, hundreds of British four-engine bombers unleashed their payload of explosives on the laboratories, but missed the target. They killed many workers, mostly prisoners of war, without causing serious damage to the facilities. For the moment, the program was safe.

An enraged Hitler immediately took countermeasures. He replaced Dornberger with a Himmler loyalist, SS General Hans Kammler, and ordered him to oversee a complete logistical restructuring of the project. An old fuel depot near Nordhausen in Thuringia, central Germany, was turned into a gigantic underground factory. Kammler, a brutal man who as a civil engineer had built several concentration camps, including Auschwitz, had the idea of using prisoners as slaves in the rocket program. The living conditions of these unfortunate people were terrible. There were many more deaths to build the missiles than victims of the missiles themselves. Obviously von Braun knew everything, but pretended not to notice. "*On a small area near the ambulance shed, inmates tortured to death by slave labor and the terror of the overseers were piling up daily*", a witness will declare. "*But, Prof. Wernher von Braun passed them so close that he was almost touching the corpses*".

Mass production of the A4 had not yet begun. The jewel needed more care. Determined to take over the project, Himmler had von Braun arrested on serious and false charges of sabotage, defeatism, and communism. The scientist was imprisoned and threatened with death. After two weeks, thanks to the good offices of the usual Dornberger and Albert Speer, Minister of Armaments, he was released with the consent of Hitler himself. The Führer, who was waiting for the winning weapon, had ordered that the head of his champion be spared.

In the meantime, however, Göring's V1s, launched from mobile ramps, had begun to torment England. But the hit rate was low, and the English had quickly learned to shoot down the clumsy rockets before they entered the Island's airspace. Nevertheless, the Aggregats continued to lose in the confrontation with the V1s, if only because they were still latent. To add insult to injury, Hitler survived an assassination attempt on July 20, 1944.

The wolf was attacked in his own hideout by his own officers, and after his miraculous escape, he became convinced of his immortality. The tragedy took on the hues of a bestial farce. Now or never, von Braun thought, and on September 7, 1944, the first A4 took off for London. It was now called the V2. Several thousand would be launched during the war, almost half of them over England and a sixth in the capital London.

A brilliant success and a real slap in the face to the vain leader of the Luftwaffe. But von Braun seemed unhappy. He would have liked to study more, test longer, improve better, and instead he had developed kamikazes that, in order to destroy the enemy, destroyed themselves. Beautiful as the dart of Zeus, economically they were a losing investment. They cost as much as a large bomber and could only be used once. Unable to explain it, the Allies began to fear that they were the vanguard of a future rocket with a nuclear warhead and took countermeasures. Now the Moon seemed really far away.

In January 1945, the German resistance was dwindling. The Red Army knocked hard on the borders of Germany, advancing rapidly, paving its way with the terrible Katyusha rockets: a tactical weapon much simpler than the V2, but very efficient. The rumors of the cruel revenge of the Soviets frightened everyone. A wind of death was blowing in Peenemünde. The engineers and technicians looked askance at the SS and spied on their movements. They feared that Himmler had ordered them to eliminate the thinking heads to prevent them from falling into enemy hands.

On January 31, Kammler instead ordered the evacuation of everyone, weapons and baggage. The destination was Nordhausen, which still seemed to be a safe place. A counter-order followed, but von Braun and Dornberger ignored it. They loaded five hundred people, tons of papers, and countless parts of the V2 onto a couple of commandeered trains, which they carefully marked with SS symbols—a safe conduct that still worked very well in the Reich—and headed for the heart of Germany. By April, even the new refuge had become unsafe, and the Peenemünde caravan was ordered to move to Bavaria, close to the American-held front.

Kammler was determined to sell the treasure of Peenemünde to the Allies in order to save his own skin. A project of conditional surrender that von Braun also shared for himself and his people because, as he later explained, *"the secret of rocketry should only get into the hands of people who read the Bible"*. To keep on living was not enough for him. He wanted to be able to continue his studies. Two trucks full of documents and projects accumulated over 13 years of work were hidden in an abandoned iron mine. Then Kammler disappeared into thin air, one of the many Nazi leaders swallowed by oblivion and left unpunished.

In the general chaos, von Braun and Dornberger found themselves together in a small village in Tyrol. They were far from desperate. *"[We lived] royally in the ski hotel on a mountain plateau [...] Hitler was dead, the war was over, an armistice was signed—and the hotel service was excellent"*. Then, with the help of Wernher's younger brother who acted as an intermediary, they surrendered to the Americans.[6] They felt safe and, much worse, in the right. *"We wouldn't have treated your atomic scientists as war criminals and I didn't expect to be treated as one. No, I wasn't afraid. It all made sense. The V2 was something we had and you didn't have. Naturally you wanted to know all about it"*.

After the surrender, Americans daringly recovered V2 parts and projects scattered throughout Yankee-occupied territory, which had to be turned over to the Russians by Allied agreement. There was no reason for fair play, said the White House. So the human prey and materials from Peenemünde were shipped to the U.S. as part of the secret operation known as Paperclip, which was designed to recover as many German scientists as possible with the intention of reusing them at home and specially to take them away from the Russians. Von Braun was one of them. Courted by the English, who had pragmatically forgotten the V2 explosions on their heads, he had chosen the United States instead.[7] He was convinced that this great nation had the interests and resources to continue his dream.

"My country has lost two wars in my youth. This time I wanted to be on the side of the winners", he explained. The Americans, in their puritanical souls, were divided on how to welcome him. Opportunism prevailed. In February 1947, he was even allowed to return to Germany to marry in the Lutheran rite his beautiful second cousin, the eighteen-year-old Maria von Quistorp, a *"porcelain lady"*, as the tabloids described her. Wernher's second life had begun, far from the agonizing cries of the forced laborers, wretches condemned to a slow and cruel death to serve the dream of glory of Baron von Braun, the Junker of the *"Gott mit uns"* (God with us).

But who was he really? Surely an engineering genius, a visionary, and an exceptional manager; but also a cynical, selfish man willing to commit any atrocity to achieve his goals. An amoral *Übermensch* (Superior Human) in the sense Friedrich Nietzsche gave this term, and a criminal of fine intelligence?

---

[6] A curiosity. In the photo of the capture, von Braun shows a conspicuous arm in a sling. He had broken it a few months earlier when his driver fell asleep and drove his car off the road. After refusing a cast, the bone had healed poorly and the limb had to be re-break and properly immobilized.

[7] Dornberger, who escaped a death sentence in England, also emigrated to the U.S., but was excluded from the government's recycling of ex-Nazis. He worked as a private citizen in the aircraft factory that would produce most of the helicopters used by American troops in Vietnam. It really is true that a leopard can't change his spots!

Hard to say. Even if thousands are shades of black, how should we judge those whose sublime minds, worthy of the Nobel Prize, gave birth to the bombs of Hiroshima and Nagasaki? Certainly, the memory of Wernher von Braun is burdened by his association with the SS and his indifference to the inhuman atrocities of Nazi fascism, while the triumphs accumulated in the conquest of space shine forth. What a difference with his greatest opponent from behind the Iron Curtain, the "chief designer": a short, stocky man with his head slumped on his shoulders, his eyes lively and sweet, and "*a relentless determination—almost like a disease*", a skeptical and pessimistic womanizer, aggressive and paternal, who would have brought his beloved homeland, the Soviet Union, very close to the Moon.

The Unnamed

Andrea—"Unhappy is the land that has no heroes".
Galileo—"No, Andrea: unhappy the land that needs heroes".
Bertolt Brecht
And our efforts are so rewarded,
For, having overcome lawlessness and darkness,
We have forged fiery wings
For our country and our age!
Monument to the Conquerors of Space, Moscow

On Thursday, October 3, 1957, in a site 200 km east of the Aral Sea in the semi-desert steppe of Kazakhstan, an icy Siberian wind blew relentlessly. It was bitterly cold, considering it was only the beginning of fall. Inside the imposing cosmodrome, the tension was palpable. A big hit was brewing, the kind that leaves an opponent stunned: the launch of an artificial satellite! A record and a first for the USSR and for the entire world, to be used in the race against the United States, and a political bonus for Nikita Khrushchev (1874–1971), secretary of the Soviet Communist Party (CPSU—*Kommunističeskaja Partija Sovetskogo Sojuza*), threatened by an internal frond. The harshest conservatives had not forgiven him[1] for the attack on Stalin's memory and criticized his clumsy handling of the Hungarian crisis and the war launched

---

[1] Vyacheslav Molotov, a man for all seasons, said of him: *"Khrushchev knew as much about matters of theory as a shoemaker. He was a real foe of Marxism-Leninism, a real enemy of communist revolution, a covert, cunning, skillfully camouflaged enemy [...] The thing is that he reflected the spirit of the overwhelming majority".*

M. Capaccioli, *Red Moon*, Springer Praxis Books, https://doi.org/10.1007/978-3-031-54760-7_5

by the French and the British against Russia's friend Gamal Abdel Nasser. A conflict that reeked of oil and had brought the whole world to the brink of the abyss.

The cosmodrome was brand new.[2] The construction of the launch pad had been completed in April, just in time for the maiden flight of the Soviets' first intercontinental ballistic missile (ICBM). A *tour de force* initiated in March 1954 by a decree of the USSR Ministry of Defense to find a suitable site for testing and launching the ICBMs, balancing the various requirements of accessibility, economy, security, and secrecy. After a year of investigations, the preference had fallen on a semi-desert area of the steppe near the Tyuratam railway station, on the line that goes from Kazansky terminal in Moscow to Tashkent in Uzbekistan. The favorable latitude to best exploit the Earth's diurnal motion as an assist to the launches and the very low local population density, which minimized the risks of news leaks and dangers to civilians in case of an accident, played in its favor. The flatness of the steppe also facilitated radio transmissions among ground control stations located within 250 km of the base to triangulate the position and speed of the spacecraft. These stations were the counterpart of the vast network that had been built on Soviet territory since the early days of the space race to provide two-way communication with spacecraft: to send commands and to collect and process information from orbit.

The construction work, mostly carried out by military engineering units, had disrupted the lives of the few inhabitants of the nearby village of Lenisk. From the nothingness of a thousand-year-old history of solitude, tens of thousands of workers and technicians, theories of endless freight trains loaded with materials, and many armed soldiers to guard had materialized in just a few months. And then the frequent explosions of the mines to dig foundations and deep trenches, the dust, and the new skyline of a corner of the steppe, made up of tall and complicated towers and sheds, illuminated even at night.

The Soviet media referred to the site of the space station as Baikonur, the name of a city very far away, 400 km from the cosmodrome in a northeasterly direction. Probably a ploy to mislead the investigations of the American intelligence, the most curious because it served the only real competitor of the USSR both in nuclear weapons and in the space race. In fact, the site remained so well covered that the CIA had to resort to high-altitude aerial espionage to find it. A clumsily handled necessity that would cause a serious international crisis in 1960 when a U2 stratospheric plane, sent specifically to

---

[2] The first and still the largest in the world. After the collapse of the USSR, the area of the cosmodrome, a square of 90 km on each side, became the territory of an independent republic; however, an international agreement assigns its use to the Russian Federation until 2050.

photograph the cosmodrome, was shot down by a SAM missile and the pilot, Lieutenant Gary Powers, was caught red-handed. In any case, the consistent use of Baikonur to indicate the Soviet launch site eventually took root and became the official name in 1995, when Boris Yeltsin gave it to Leninsk, the village where the cosmodrome staff had grown into a city.

The satellite launch was scheduled for three days later, on the evening of October 6 at 10:30 Moscow time, or half past midnight local time. A night window is useful both to take full advantage of the Earth's rotation and gain free speed for the vector, and to protect the operation from aerial espionage. The exceptional event had attracted illustrious guests, politicians, and military personnel to the cosmodrome. For everyone, the watchword was to keep silence, especially if things should go wrong. The world would be informed only in case of success, and then only through the Party's communication organs. A standard procedure in all dictatorships, which in the USSR was handled by Tass; the agency that, according to a decree of the Presidium of the Supreme Soviet, "*had the exclusive right to collect and distribute information outside the Soviet Union, as well as the right to distribute foreign and internal information within the Soviet Union and to manage the press agencies of the Soviet republics*".

In the group of big shots wandering around the compound, among the olive green uniforms of the generals and the fur hats of the civilians, there was a fifty-year-old man bundled up in a wide-brimmed black leather coat. He had a nice round face and a voice that was now persuasive and now thunderous. The strangers addressed him as *glavny konstruktor*, chief designer, to which he nodded with a hint of a smile to encourage the speaker to continue. But then he barely listened; his mind was elsewhere. He felt the weight of those days on him like a boulder. His had been the promise to Khrushchev to put an artificial Moon into orbit around the Earth before anyone else, as proof of the Soviet Union's superior technological capabilities. And, as at the Olympics, communists around the world eagerly awaited a podium to brag about.

The feat was part of the "sporting" side of the fierce clash between two ideologies and two different economic systems. For Khrushchev, it meant prestige abroad and greater solidity within the party and state apparatus. For the chief designer, it was the realization of a long-awaited dream, paid for in advance at a very high price. The moment of truth was at hand. While he was pondering the possible problems of the launch, growling some orders and reprimanding some subordinates with a stream of insults, he received a telegram that made him turn pale. The 007s at the Soviet Embassy in

Washington had sniffed out that the Yankees were about to launch. But when?

In the grip of growing anxiety, the *glavny konstruktor* demanded to speak to Moscow immediately, but the call only increased his stress. The danger of a last-minute American overtaking seemed real. In fact, the secret service agents were mistaken, but that would only be known later. In the meantime, the cards had been dealt, and in total ignorance of the opponent's hand, there was nothing left to do but to play ahead and launch the next day. Taking responsibility for this grave decision, with the same courage with which Nelson disobeyed his commander at Trafalgar and succeeded in scattering Napoleon's fleet, the Man in Black gave the order and the dance began. The rocket that would serve as the slingshot for the satellite was called the R-7 *Semyorka*, a word that means "seven" in Russian, and it was now sleeping in a hangar. Its various parts had arrived by train from factories mostly in Moscow, Leningrad, and Kharkiv regions. Pieces of a high-tech puzzle that were then assembled in the technical area of the Baikonur Cosmodrome.

An almost identical copy of the R-7 had been successfully tested a month and a half earlier. The giant, loaded with a dummy nuclear warhead, had majestically taken off from the *ploshchadka* (site) no. 1 and had meekly reached its target on the Kamchatka Peninsula. Tass had reported the success—marred by a minor incident at the end of the flight and omitted, as usual, in the press release—with a laconic: "*On August 27, 1957, the USSR realized the first multistage intercontinental ballistic missile*". A word to the wise is enough! Turning this formidable war machine into a taxi for the *Sputniks* required only two quick modifications. A new nose, shorter and lighter than the one designed to house and protect the atomic bomb, and an overhaul of the fuel to increase thrust and reach the minimum speed required to launch a small 84-kg sphere into low Earth orbit.

A clarification is needed at this point. What is the difference between firing a cannonball and launching an object into orbit? In the first case, the projectile spends the vertical component of the velocity it impulsively acquired at the time of firing to gain altitude, and then, after exhausting its thrust, it falls back to the ground (assuming, of course, that the bang did not cause the projectile to escape the Earth's gravitational field; but that requires a very high initial velocity, 11.2 km/s). Our Moon, instead, does not fall on our heads because the Earth's gravitational pull is exactly balanced by the centripetal force due to a velocity that not only has to have the right value but, in the simple case of circular orbits, must also be just tangential, i.e., perpendicular to the orbital radius. This equilibrium condition holds for any other radius, larger or smaller than that of the Moon's orbit, provided that the altitude is high enough so that the braking effect of atmospheric friction can be neglected.

More generally, if you want to put something into orbit, you must arrange things in such a way that, once the object has reached the desired altitude, its velocity has both the value and direction required by equilibrium in the wanted trajectory; and this cannot be done through a single impulse. For instance, at 223 km of altitude, *Sputnik* needed to travel al 7.8 km per second almost perpendicularly to the vertical. This is a respectable speed, which would allow us to go from Rome to Milan in just one minute as the crow flies, and which the R-7 was able to provide thanks to four additional burners connected to the second stage, to be jettisoned after use.

Let's go back to Tyuratam where, according to the new schedule, there were just over 24 h left before the launch. The chief designer ordered his men to immediately move the rocket from the hangar to the launch pad. A short walk, at human pace, to reach a massive reinforced concrete balcony with a hole in the center allowing the venting of flaming gases into a large basin 35 m below. Once in place, the giant was lifted by crane and hooked up to mobile service structures that enveloped it like the petals of a carnivorous plant. Teams of white-coated engineers began the final checks in light of photoelectric cells as the liquid oxygen, kerosene, and amorphous gases were loaded. It was a long and dangerous operation in which the slightest mistake could be fatal for both the thoroughbred and its keepers.

The deputy defense minister and head of the Strategic Rocket Forces, the highly decorated general Mitrofan Nedelin (1902–1960), watched the operations with a dark face. He had been the promoter of the program that had given birth to the R-7. He believed that an intercontinental ballistic missile might be the best answer to the challenge posed by the Americans with their B-52 Stratofortresses. These long-range strategic bombers, giant subsonic jets produced by Boeing, were designed to deliver heavy nuclear warheads anywhere in the world, either by in-flight refueling or by takeoff from allied bases near Soviet territory.

Nedelin could touch on hand the weakness of his project. The ability to carry a large bomb across the Atlantic was not sufficient to qualify the R-7 as an effective tactical defense weapon, since the launcher took a full day to become operational. The chief designer's creature behaved like a boxer who has a deadly hook but uses it in slow motion. The enemy, the high-ranking Red Army officer thought with great disappointment, would have plenty of time to wipe out the R-7s in their parking lots while they were still sucking fuel from their giant feeding bottles. At least for the time being, it was technically impossible to keep these heavenly torpedoes awake and ready to pounce on the enemy with full tanks. A big flop on the military level. The *glavny konstruktor* understood the problem, but for the moment he had other things to worry about. After all, in his heart he cared relatively little about the military and its needs. What he wanted was to conquer outer space, and he dreamed of traveling to other worlds.

Eventually, night fell again. As the launch time approached, the area was evacuated. There was a real risk that the rocket would explode at takeoff. The authorities and technicians who were essential for the maneuvers took refuge in the command bunker. The environment was austere, a thousand miles away from the huge rooms, glittering with LEDs and screens, that the media show us today; almost like theaters, where there is also room for an audience of guests. At Baikonur, luxury was a word without meaning, and even the essentials were often broken and in poor condition. In Machiavelli's words, what mattered was not the means but the goal.

The Russians at that time did not use the countdown technique already in vogue in America to sequence and monitor the various stages of preparation for launch. They launched when the *glavny konstruktor* said so. And he said it, even though he had been informed that one of the boosters had failed to complete fueling. But stopping would have meant losing the launch window, postponing it for another day, and perhaps being overtaken by the hated capitalists at the last minute. The risk had to be taken, whatever the cost.

The sequence of the various steps of the launch protocol began, including a low-regime preheat of the engines. The giant puffed and spat flames. The hellish scene could be seen well through one of the two periscopes in the bunker. The noise was deafening, like the roar of a giant wounded beast. Finally, full power was given, and the R-7, a 34-m-high, 380-ton obelisk, after a programmed pause of a few seconds, began to lift off faster and faster until it became a faint glow in the blackness of the sky. Silent in a white lab coat that fit him too tightly, the chief designer was extremely tense. His career and all his hopes were on the line. He was carefully checking the readings of the various instruments, and, if someone raised the voice or showed signs of nervousness, he was immediately on alert to see what was happening.

Understandable concern, but unnecessary under the circumstances. The rocket's ascent, followed by a chain of tracking stations scattered across the vast country, was going well. The fuel problem did not seem to have any serious consequences. Then, loud and clear, came the beep of *Sputnik*, which, ejected from its carrier, had set off on its celestial trail: an orbit inclined 65° to the equator, with a perigee at an altitude of 215 km and an apogee at 940 km, covered in 96 min. Half an hour passed and the signal disappeared—the satellite had dropped below the horizon of the stations on Soviet territory—to reappear only after an hour, as expected. Joy exploded in the bunker. Everyone hugged each other. Perhaps a few bottles of vodka and Armenian champagne materialized, and a few tins of black caviar. For a moment the *glavny konstruktor* forgot the terrible injustice he had suffered and passionately kissed the chief engineer, Valentin Glushko.

The next day, the first of the space era, the chief designer called an informal meeting of all the scientific and technical personnel and improvised an unusually flowery speech:

Dear comrades! Today, that which the best human minds have dreamed of has taken place! The prophetic words of Konstantin Eduardovich Tsiolkovsky that mankind would not always remain on the Earth have come to pass. Today the first artificial satellite in the world has been injected into orbit around the Earth. With its injection into orbit, the conquest of space has begun. And the first country to open the road to outer space has been our country – the land of the Soviets! Allow me to congratulate all of you on this historic occasion. And allow me to especially thank all the junior specialists, technicians, engineers, and designers who took part in the preparation of the [rocket] and the satellite, for their titanic labor [...] Once more I give to you a hearty Russian thanks!

Then the key project leaders took a plane to Moscow. During the night flight, the pilot informed his passengers that radio stations around the world were buzzing with news: "*All you can hear are words like satellite and Russia in all languages*", he told them. The chief designer gloated, but with a wry smile he simply replied, with a phlegm that would make an English lord envious: "*It seems we have caused quite a stir*".

News of the resounding success had been readily communicated to the party secretary, who immediately wanted to congratulate the authors. The night of the launch Khrushchev was in Kiev, a city he loved particularly, to attend an evening reception at the Mariyinsky Palace, rebuilt from the ruins of the war. The vodka flowed like rivers there. At eleven, one of his assistants told him there was an urgent call from the Kremlin. Nikita, who did not disdain to lift his elbow, left the room somewhat unsteadily. When he reappeared in public, "*he had a huge smile*", his son Sergei later recalled in his memoirs.

He finally couldn't resist saying [to the Ukrainian officials]: "[Tovarishchi], I can tell you some very pleasant and important news. Korolev just called (at this point he acquired a secretive look). He's one of our missile designers. Remember not to mention his name – it's classified. Korolev has just reported that today, a little while ago, an artificial satellite of the Earth was launched.

The word "satellite"[3] meant nothing to any of the listeners, simple party boyars who heard Korolev's name for the first time. Surely they all pretended to have immediately forgotten what should not be remembered. Stalin had trained them very well to obey blindly; but probably someone in his heart wondered who the hell this unmentionable guy was.

Sergei Pavlovich Korolev (1907–1966) was born in Zhytomyr, a historic city in Western Ukraine. His father, Pavel Yakovlevich, a teacher of Russian language and literature of humble peasant origins, came from Belarus. His mother, Maria Nikolayevna, instead belonged to a wealthy Ukrainian family of fur traders from the town of Nizhyn and boasted of genuine Cossack descent. The blood cocktail was completed by Greek and Polish ancestors. Pavel and Maria's marriage did not last long. The woman could hardly endure her husband's meager existence. Sergei was just three years old when they divorced. The child was taken to live with his maternal grandparents in Nizhyn, where he grew up in great loneliness. He had few friends. His brilliant intelligence and *"stubborn, obstinate, and quarrelsome"* character earned him the dislike and hostility of his peers, who were jealous of his scholastic achievements. He also suffered from emotional deficiencies. Maria had resumed her education and was often absent from Kiev. In 1916, she remarried to a German-educated electrical engineer who had just gotten a job with the southeastern regional railroad. The new family moved to Odessa the following year. The stepfather, a cultivated and kind man, proved to be a real father and cultural reference for young Sergei, to whom he also passed on his love of science and technology.

These were difficult times. Tsarist Russia, which had been at war for three years, was collapsing. Sergei's arrival in his new home on the Black Sea coast practically coincided with the outbreak of the October Revolution. Odessa, which had already suffered the turmoil of the 1905 workers' revolt, supported by the sailors of the battleship Potemkin (depicted in a famous movie by Sergei Eisenstein), was occupied by the Central Ukrainian Committee, then by the French army, by the white army of nostalgics supported by the West, and by that of the red revolutionaries.

The danger and violence ended in 1920, when the city and its territory were definitively annexed to the Soviet Union by the Bolsheviks. In the meantime, the schools had remained closed and Sergei was forced to study at

---

[3] The term comes from the Latin *"satelles"*, meaning servant or bodyguard. It was first introduced into astronomy by Johannes Kepler.

home. During this difficult period, he avidly read everything from mathematics to German poetry, from engineering manuals on material stress to the writings of a French anarchist expert on balloons, "*drinking it all and keeping it in mind*". In 1922, he passed with flying colors the final exams of the construction trade school, where he met Ksenia Vincentini, a girl "*slender, very pretty, with a braid that hung below her waist and big eyes*". In 1931, she would become his wife and the mother of their only daughter, Natalia. Two years later, Sergei went north to enroll in the Kiev Polytechnic Institute. It was a retreat. He would have liked to go to the Moscow Military Academy, named after Nikolai Zhukovsky (1847–1921), the man who was the first in the world to mathematically solve many of the problems of flight, opening the way to a new branch of engineering. He already had clear ideas.

In fact, the proximity of a seaplane squadron of the Odessa Coast Guard to his home had long since ignited his passion for flying. As a boy, he often swam long distances to enjoy their acrobatics at close range. He even built a glider in his spare time, and at the age of sixteen was one of the first to join the Friends of Air Fleet Society. He often flew, with courage and great coolness in dangerous situations. Once he fell into the sea while climbing on the wing of a broken biplane to check the oil level. A 10-m dive. He escaped thanks to his outstanding athletic qualities. A physical form maintained by swimming and gymnastics, which fifteen years later in Siberia would have allowed him to survive the cold and hardships.

In the capital, the beautiful Kiev, situated on the hills overlooking the grand Dniepr, he found other companions with whom to share his growing passion. Another two years passed, and with another leap Sergei finally moved to Moscow, the capital of a great country again under construction, to attend a prestigious engineering school, the Bauman Moscow Higher Technical School, where he graduated in 1929 with an exceptional mentor, the famous aircraft designer Andrei Tupolev (1888–1972). He was twenty-two years old, with a love in his heart for his Ksenia (who still did not reciprocate) and an academic title that could be used anywhere, and a now mature passion for flying.

For a while he worked in an aircraft factory. Then, together with his stepfather he designed and built an acrobatic glider that was praised by none other than the great Sergei Ilyushin (1894–1977), another giant of aeronautical design. Finally, in 1931, his master Tupolev called him to the Central Aerodynamics Institute, the *Tsentral'nyy Aerogidrodinamicheskiy Institut* (TsAGI), with the task of developing an autopilot for the heavy bomber project he was working on. The institute had been founded in Moscow in 1918 by Zhukovsky, the father of Russian aviation, three years before his death.

How many geniuses, one might speculate, were concentrated in Great Russia between the end of the nineteenth and the beginning of the twentieth century! That's not so surprising. Every civilization must approach its sunset in order to reach its own cultural maturity. It's a matter of inertia. Power decays first and faster than ingenuity. Imperial Russia had completed its life cycle, and a new form of society was maturing on the ruins of the one that had disappeared. New ideals that could draw on a rich heritage of knowledge.

Since 1927, Stalin had been the absolute ruler of a country the size of a continent, with widespread poverty and backwardness, but with high levels of intelligence and competence. The dictator had many priorities. Rationalize agriculture to defeat hunger; industrialize production to compete with the Western system; expand services, including schools, to eradicate illiteracy and make use of brains that would otherwise be wasted; and above all, defend himself, maintain power, and promote his own image. Unfortunately, the immense energy, courage, and political instinct of this man, who was born a priest and became an armed rebel, did not meet with an equally great pathos. He admitted it himself, if it is true that in 1907, in despair over the death of Nadezhda, his beloved first wife, he confessed to a friend: "*This creature softened my heart of stone. She died, and with her died my last warm feeling for humanity*".

Sergei had not yet been infected by rocket science, although he had attended several conferences on the futuristic subject of space conquest on the advice of his teachers. He was interested in this technology mainly for its application to aircraft. Apart from Tsiolkovsky, who acted like a lone wolf, the subject was already alive and present in the USSR at the Gas Dynamics Laboratory in Leningrad (the new revolutionary name of the imperial capital of St. Petersburg), where the engineer Glushko was working—the same person we met a few pages earlier at the launch of *Sputnik* at Tyuratam. The research institute, responsible among other things for the development of the Katyusha rockets[4] that would make their deadly song heard by the enemy during the Patriotic War, had its origins in a private enterprise promoted by Lenin himself for the "*construction and welfare of a young nation of workers and peasants*".

In Moscow, the matter remained in the hands of a group of amateurs, similar to the German VfR and the American Interplanetary Society, called the Group for Investigation of Reactive Motion (GIRD—*Gruppa Isutcheniya Reaktivnovo Dvisheniya*). A sort of club for rocketry enthusiasts, it quickly

---

[4] These rockets equipped the so-called Stalin's Organ, a battery of dozens of tactical missiles, mostly transported by trucks.

took root in Leningrad, and later in Baku, Tbilisi, Belarus, and the Caucasus. Korolev began to visit the Moscow section more and more often, together with a colleague from TsAGI, the Latvian Friedrich Tsander (1887–1933).

Twenty years his senior, Tsander was one of the first proselytes of Tsiolkovsky and had already designed rocket engines and even published a long article, *Flights to Other Planets*, in 1924, full of new ideas (spaceships with reflecting mirrors and solar sails). A Baltic German, whom the history of science remembers as "*the first man in the USSR who took the necessary steps to make astronautics an applied science*". As the months passed, the two friends, now in love with Oberth and through him with Goddard, devoted much of their free time to GIRD. The result of this activity was a liquid propellant not very different from the first ones developed in Berlin.

At the same time, in Leningrad, Valentin Glushko (1908–1989) was speculating on a rocket engine with an impulse generated by electric charges accelerated by an intense potential difference, which, as he would later say, with his characteristic egocentrism, "*anticipated the state of science and technology by about thirty years*".[5] A peer of Sergei's, Valentin, was also from Odessa, the son of poor people; but although the two boys attended the same school, they never had a chance to meet. At the age of thirteen, he fell in love with space, and in 1923 he expressed his vocation in a letter to Tsiolkovsky in Kaluga. In return, the old guru, like a true crusader, asked him how strong his faith was. "*I want to devote my life to this great cause*", was the young man's emblematic answer. He would keep his promise, even at the cost of his life.

The engine of Tsander and Korolev had several problems. Finally, in 1933, it was ready for launch aboard a sleek shell. Named GIRD-X, on its maiden flight it reached an altitude of 5 km. Tsander could not rejoice. Typhoid fever had taken his life a few months earlier. As in Germany, the success of the enterprise drew the attention of the military to GIRD's activities and led to its merger with the Leningrad Gas Dynamics Laboratory to form a new entity, the Reactive Scientific-Research Institute (RNII—*Reaktivnyy Nauchno-Issledovatel'skiy Institute*) based in Moscow. A decree of the Council of Labor and Defense placed it under the jurisdiction of the People's Commissariat of Defense Industry and entrusted its control to an atypical Red Army general, Mikhail Tukhachevsky (1893–1937), a man of noble origins, refined manners, and great culture. He spoke several languages, played the piano brilliantly, and wrote books. More a remnant of old tsarist Russia than a classical Bolshevik. The direction was left to the military engineer Ivan Kleimenov, who was already in charge of the Gas Dynamics Laboratory. Korolev was

---

[5] This type of propulsion is currently in use on NASA's mission to the asteroid Psyche, launched in October 2023.

made his deputy. A quick promotion, but of short duration. Having clashed with his boss, Sergei was soon demoted. An act of persecution that would save his skin three years later.

Then came the season of great terror. In the late 1920s, Stalin, the archetype of the strong man (his battle name means "Man of Steel"), the bloodthirsty leader, and the charismatic dictator, had launched a five-year program of state renewal and power management based on the most brutal violence. His attention was focused on both the sickle and the hammer. He had introduced forced collectivization and the mechanization of agriculture, strengthened heavy industry, and promoted the birth of new cities to house the workers, new schools, and universities to provide the minds with which to realize his project. Those who opposed were arrested, deported, or killed. The small landowners of peasant Ukraine, who had tried to resist the communist administration of the land between 1932 and 1933, were literally starved to death by a famine invented on the drawing board, the *Holodomor*, the painful memory of which still lingers. "*A single death is a tragedy*", Stalin cynically commented; "*a million deaths is a statistic*". And he acted accordingly. The British government had used the same harshness during the Great Famine of 1845–1852 in Ireland, refusing to stop the massive export of food from the island.

Between 1936 and 1938, the dictator's blind sword fell on almost the entire old Bolshevik guard. Obsessed with conspiracies, Stalin wanted to purge the same CPSU, believing that the field had to be plowed in order to sow the new. The heads of 25% of the officers of the Red Army fell. A massacre that would have serious repercussions on the initial conduct of the war against the Nazis due to the lack of leadership. The dictate for his courts was to overturn the ancient Justinian rule of "*in dubio pro reo*" (in doubt, rule for the accused). A minimum of suspicion was enough to get into big trouble. This is exactly what happened to Glushko and Korolev.

In June 1937, as Hitler and Mussolini were sharpening their claws against the Spanish Communists, Tass announced that Marshal Tuchayevsky, a prominent figure in the Bolshevik Party, had been arrested and shot on the usual charge of "*enemy of the people*" for being a "*spy for the Germans*". His mother, sister, and two brothers were also killed, *ad abundantiam*. In reality, the evidence against him had been artfully concocted on Stalin's orders by the People's Commissariat for Internal Affairs (NKVD—*Narodný Komissariat Vnutrennih Del*), the Soviet secret police agency in charge of internal security and labor camps.

Tuchayevsky's dealings with the RNII led to the disinfestation of the Institute, with the elimination of all senior staff, including Glushko and Korolev.

Perhaps another trigger for the arrests was an accusation by a colleague who wanted to get rid of the competition; or possibly, it was said, each of those arrested in a chain called on another, hoping to obtain some credit and a better treatment. The logic behind these actions may have been "bring us a friend and you get a discount".

On June 27, 1938, in the middle of the night, four people, two NKVD agents and two witnesses, burst into Korolev's house and ordered him to gather some of his things and follow them. When Sergei asked for explanations, he was told that he had squandered public money. They took him to the infamous Lefortovo prison in the homonymous district of central Moscow, where he was interrogated and beaten. When he pleaded for a glass of water, a policeman hit him in the face with a jug, breaking his jaw. A nightmare! He was then brought before a judge who ordered him to confess his guilt. "*What crimes?*", muttered the shocked defendant. "*The ones listed in the paper you were given*", the judge replied. "*But I have done nothing*", Sergei tried to defend himself. "*All of you criminals claim to be innocent. You are guilty! I condemn you to 10 years of hard labor. Next*". It looks like a scene from a Jacobin court. The desperate appeals of a distraught Sergei and his mother to Stalin himself were in vain.

Korolev was first imprisoned and then sent to the Kolyma region, in the far northeast of Russia, above the Arctic Circle. It was winter, and the temperature in the tundra dropped to 30 degrees below zero. The *gulag* was located at an open pit gold mine. The prisoners cut down trees, dug holes, and pulled carts. Food was scarce, two soups a day and 80 g of bread; clothing was completely inadequate. People died in droves. Sergei lost almost all his teeth to scurvy. Then, almost unexpectedly, this horror ended for him. He was sent to a labor camp for the intellectual elite, the *sharashka* OKB-29 (*Opytno Konstruktorskoe Bjuro*—Experiment and Design Bureau), and his sentence was reduced to eight years. What had happened?

Stalin, who always had his ears pricked to catch his opponents' moves and could rely on the help of many communist sympathizers scattered around the world, had learned of the progress made by von Braun's team at Peenemünde. Impressed, he ordered his most loyal compatriot, Lavrentiy Beria, now at the head of the NKVD, to take charge of the matter. A well-placed choice, since this man, treacherous, vicious, and cruel, was an extraordinary organizer. It was necessary to recover as many scientists as possible and put them to work under the direction of Tupolev, who had also been arrested in the meantime, with the usual refrain of sabotage, counterrevolutionary activities, and espionage.

Beria established the Tupolev *sharashka* in Moscow and gathered the best aviation and rocket specialists from the *gulag*. Among them were two convicts, Korolev and Glushko, who found themselves working together, although they had little sympathy for each other. Sergei even thought it was Valentin who denounced him to protect himself. The living conditions in the VIP concentration camp were more humane and, above all, heavy work was replaced by intellectual activity, much less tiring and much more rewarding. "*All animals are equal, but some animals are more equal than others*", George Orwell wisely said.

In the *sharashka*, scientists were scientists again, serving with devotion the homeland that had torn them apart and yet they continued to love desperately. A loyalty that would reach new heights at Korolev. The winds of war suggested concentrating on the products that would be needed to win: the aircraft. So the rockets were shelved for the time being, but not their engines, a subject in which Glushko was a great expert, because they could be used to make planes faster and more powerful. In the race to the stratosphere, the German competitors in Europe were now left to compete with themselves at Peenemünde.

On May 27, 1941, the winds of war became a storm. Hitler, disappointed with the results of the Battle of Britain, had ordered the invasion of Russia. He wanted a spectacular and quick success. So he put all the power and effectiveness of his land and air forces into play. A deployment of troops, on two fronts, that had never been seen before. Unprepared, Stalin had called the people to arms, even proposing an armistice to the Pope of Moscow, so that everyone, but everyone, would participate in the patriotic war by giving their all.

Just as Napoleon's *Grande Armée* had done 130 years earlier, the highly mechanized troops of the Führer quickly arrived at the gates of Moscow and Leningrad. There, one step away from temporary triumph, they were stopped by the megalomania of a strategic design that continued to ignore the lessons of history, by the desperate courage of soldiers and civilians, and by the most reliable of Russian servers, "General Winter". By the end of the year, the initial German momentum had been halted. But the real turning point came with the battle for the city of Stalingrad, a strategic gateway to the oil of the Caucasus, which Germany desperately needed. A bloody battle that lasted from mid-1942 until February of the following year. The Soviets won, with acts of heroism that the Russian people cannot forget, even 80 years later.

By early spring 1945, the Red Army had reached the eastern borders of Germany. The end of the Third Reich was only days away. Having overcome the great fear, Stalin now considered how to use the victory. His generals were ordered to plunder Nazi technology. Immediately after the Soviets had reached and conquered the scientific centers and industrial regions of Germany, a commission of experts was to proceed with the removal, safe-keeping, and shipment to Moscow of consolidated or experimental war equipment, industrial machinery and research apparatus, scientific libraries, and archives. As with the similar activity of the Americans, Operation *Osoaviakhim* (acronym for Society for the Promotion of Defense, Aviation, and Chemistry) included the forced transfer to the USSR of more than 2500 specialists, scientists, and technicians of the Führer together with their families.

The V2s were a tempting morsel that von Braun served the Yanks on a silver platter. But the Russians arrived first at Peenemünde, where something remained. Moreover, not all German technicians had chosen America. A task force of Red Army specialists was needed to collect the many crumbs and recombine them into a coherent picture. This exigence finally opened the doors of the soft prison for Glushko and Korolev, whose *sharashka*, when the Wehrmacht appeared at the walls of Moscow, had been transferred to Kazan as proof that this "dirty dozen" of convicts was now considered a value to be protected. Both scientists were effectively amnestied and sent to Germany. Sergei with the new rank of Colonel of the Red Army. He was alive, toothless, his health now undermined, abandoned by his wife Xenia, who had repudi-ated him and was now a doctor in Ukraine. Yet, amazing as it may seem, his faith in communism was intact, as was his thirst for space.

The booty was far from negligible. Hundreds of technicians who had worked on the design and mass production of the V2s were tracked down and transferred to the USSR, along with countless parts of the rockets that, with a clever operation of reverse engineering, made it possible to build and test a faithful Soviet replica of von Braun's rockets. While in America the debate continued about who should do what, in the arrogant belief that they had no rivals, the Soviets were making giant strides, recovering their own talents from the *gulag* with such a speed that the German technicians soon became useless.

Far from being reassured about the true intentions of the West, Stalin wanted his missiles to strengthen the Red Army in a defensive role, but also to support the spread of communist ideology. His idea of force was huge masses of soldiers equipped with traditional weapons, albeit adapted to technological progress. Primarily steel, as in his own combat name. But his instinct made him understand the importance of innovation, and his will was enough to put wings on the feet of those who had to reproduce it. As in the case of the atomic bomb.

Even before the war, the German project for a device to harness the energy contained in atomic nuclei was something of an open secret. The large uranium purification and plutonium production plants were too conspicuous to escape espionage, so much so that during the conflict they were systematically attacked by the Allies. From the point of view of intelligence, Stalin enjoyed a privileged situation, being able to rely on first-hand information stolen by some Western scientists of pure communist faith. He was therefore well aware of all, but he had underestimated the problem until August 1945, when the annihilation of the Japanese cities of Hiroshima and Nagasaki confronted him with the immense power of the new weapon. The bomb had upset the balance of the world. Truman went out of his way to emphasize this at the Cecilienhof meeting, in the mistaken belief, not shared by the bellicose Churchill, that U.S. supremacy could serve to secure a lasting peace.

Stalin immediately devised a plan to bridge the extremely dangerous gap. A strategy similar to that of the Manhattan Project: form a team of the country's best scientists and provide them with all the economic resources and safe conduct necessary to achieve the result in a short time. To head the fast-track program, the General Secretary chose the loyal Beria, a man of subtle persuasion, who was given full authority and an unlimited budget. "*He was the personification of evil in modern Russian history, [but] he also possessed great energy and ability to work, […] intelligence, willpower, and determination*". The physicist Igor Kurchatov (1903–1960), who had worked on this field before the war, was appointed scientific director, along with Pyotr Kapitza (1894–1984). But the latter withdrew after some time for moral reasons and was placed under house arrest.

Working day and night under the constant pressure of the NKVD chief, who never threatened in vain, and benefiting from the information of agents[6] who knew the secrets of the Manhattan Project, Russian scientists and technicians quickly arrived at the port. The first Soviet fission bomb was successfully detonated in Kazakhstan on August 29, 1949. Its dark mushroom cloud announced to the world the end of the American monopoly and the beginning of a new era in history, with the establishment of a state of permanent and cold war. It replaced the hot version, made deadly for all by the power unleashed by the atom. But having the bomb was only one part of the problem. It had to be delivered to the target. So Stalin set up a second task force to build airplanes and rockets suitable for the purpose. At the top of the pyramid he placed Georgy Malenkov (1902–1988), an experienced

---

[6] In particular, the German-born British physicist Klaus Fuchs confessed to passing the atomic bomb designs to the Russians. It is believed that without his help, Stalin would have had to wait another two years to produce his own nuclear deterrent and even settle the score with the U.S.

commander in the Great Patriotic War and a faithful executor of his boss's most heinous orders. They included the attack on Marshal Zhukov, falsely accused of Bonapartism,[7] and the persecution of Leningrad's *intelligentsia*, which, according to the dictator, reeked of dangerous nostalgia.

The road to the bomb and its deployment had been paved a few years earlier, in May 1946, with the creation of the Scientific-Research Institute No. 88 (*Nauchno-Issledovatelskiy Institut* or simply NII-88) on the outskirts of Moscow, in old and dilapidated buildings. Considered a national priority, its mission was to manage all engineering aspects of Soviet industry for the development of long-range missiles. It was divided into several bureaus (departments) and placed under the strict control of the military for security reasons and because most of its funding came from the massive budget allocated for defense. In short, the Red Army served as the NII-88's customer, guarantor, and bodyguard.

Upon his return from the long mission in Germany, Korolev, still formally a pardoned criminal and therefore under the control of the secret police, was decorated—Russians love medals—and placed at the head of the NII-88 Special Design Bureau, in a whirlwind of contradictions of a system riddled with bureaucracy. Under him, a large group of Soviet and German technicians had found hospitality in a complex built on the Gorodomlya Island in Lake Selinger, Tver *Oblast*, 300 km northwest of Moscow, surrounded by barbed wire and protected day and night by armed guards and dogs. Everything had to be top secret, including the name of the *glavny konstruktor*. For his part, while remaining undercover, he managed to establish a reputation as a skilled engineer and a rock-solid manager, capable, loyal, and very commanding.

It was imperative to proceed step by step. Korolev was assigned to study a series of ballistic missiles for the armed forces. The only previous example to draw inspiration from was the V2, which was promptly reproduced thanks to information acquired from the Germans. The first Soviet versions of the German flying bomb, called R-1 (where R stands for *rakieta*, i.e., rocket), took off from a remote village in the steppes north of the Caspian Sea in the fall of 1947: impressive but now obsolete machines, relatively reliable and with an insufficient payload. The growing needs of the military required much more power and therefore unsustainable amounts of fuel.

To overcome the impasse, a close associate of Korolev, Mikhail Tikhonravov, tried to push the concept of rocket trains, originally conceived by the mythical Tsiolkovsky, through the viscous military bureaucracy, using the "wagons" in parallel. But how? And who was this shy genius, without whom

---

[7] In Marxist theory, this term refers to those leaders who, having risen to prominence through revolutionaries, favor the capitalist elite without sharing power with them.

the explosive energy of the chief designer would not have been enough to accomplish the feats that in the 1960s gave the USSR the palm of victory in the space race? "*Together*", Sergei Khrushchev would later say of the two, "*they created the critical mass that astonished the world*".

Tikhonravov was born in 1900, when Great Russia was still ruled by the Tsar, to a family of teachers in Vladimir, the old Russian capital 200 km east of Moscow. As a boy, he learned languages and classical culture from his parents, acquiring a style and a *habitus* that he would never lose. It was this humanistic upbringing that led him to coin the term "cosmonaut" many years later.[8] Growing up, like many of his peers, he developed a passion for flying. He decided to graduate from the Zhukovsky Air Force Academy (now Zhukovsky-Gagarin) based in Voronezh, in southern Russia, and then work in the nascent aviation industry. In his spare time, he spied on the flight of birds and insects, trying to glean some secret useful for his glider projects: a lifelong hobby.

He discovered space when he learned that his old friend Korolev, together with Tsander, was trying to equip a glider with a rudimentary rocket. He became one of the founders of GIRD. His was the liquid-fueled rocket that reached an altitude of 400 m in August 1933, attracting the attention of the army. For launch, the rudimentary device had to be quickly transported by tractor to the outskirts of Moscow before the liquid oxygen evaporated. Mikhail was not present at the test because he was on a river trip to recover from overwork. Test passed, Korolev telegraphed him to announce the success of the venture, which, seven years late, replicated the pioneering flight of Goddard's rocket. With great honesty, the future tsar of Soviet space gave him all the credit. Tikhonravov, in turn, protested when a memorial stone bearing only his name was erected at the launch site in the 1960s, demanding that the entire team be credited. Hats off to both!

Then came the RNII-88 Institute and later the Stalinist purges. Perhaps because of his minimalist attitude, Tikhonravov managed to escape the dictator's axe, which instead fell on the necks of many of his friends. In any case, during these dark times he always had a suitcase at the ready, his wife later confessed. During the Patriotic War, he worked on several projects, including Katyusha rockets. When peace returned, he was reassigned to the group trying to reproduce the V2s, without the benefit of a direct confrontation with their father, Wernher von Braun, as the Americans did.

---

[8] There is no difference between a cosmonaut and an astronaut. They both mean the same thing, that is, a person who travels in the celestial space, the cosmos or the realm of stars, whatever it is. Russians prefer the former, Americans the latter.

He spent the rest of his life in a Moscow suburb so secret that it was not even on the city's maps. A closed area, guarded by the military police, gray and sad. Almost a prison of body and soul, where some of the USSR's most sensitive activities related to nuclear weapons and the means to deliver them were housed. There, in the Fourth Scientific-Research Institute of the Ministry of Defense (NII-4), Tikhonravov gathered around him a team of talents. It was in this incubator that the winning idea was born.

The problem was getting the rocket engine to work in empty space. Tikhonravov asked himself: why not utilize all the thrust at launch, when the missile is close to the ground? But how? His answer was to build a set of rockets working in parallel instead of a single buster. At first, this suggestion was dismissed as *"drifting into the realm of fantasy"*. Tikhonravov went ahead anyway. His seemed to be academic exercises, expounded with the utmost caution, because they did not conform to the pragmatism of Stalin, who was interested far more in armored divisions than in science.[9]

Meanwhile, Korolev scored one success after another with his rockets. The R-2, much better than the R-1, was surpassed by the R-3, which could fly 3000 km. Glushko worked on the engines. By now, there was little German in the revamped Soviet technology. To the point that the engineers of the Reich had become a burden. For fear of espionage, they were kept further and further away from the qualifying aspects of the projects, until, by now completely unmotivated, in the very early 1950s, they were sent back to East Germany.

To access military secrets, one had to be above suspicion. For this reason, Korolev applied for a Communist Party card in 1953, which was promptly granted, although he still had a criminal record. He would not be rehabilitated until 1957, along with many others, after Khrushchev's de-Stalinization had begun. A new course of socialism made possible by the death of the dictator.

At dawn on March 5, 1953, Stalin was found dying in his room under unclear circumstances. He was 74 years old. A stroke, it was said, but some suggested it was the result of a plot orchestrated by the boyars. That evening, the Russian people wept desperately, but in the palace, in the country, and around the world, more than a few breathed a sigh of relief. With the death of the leader and the elimination of his mastiff, Lavrentiy Beria, arrested and then made to disappear under still unclear circumstances, the intellectuals' resignation to the oppression and paranoia of Stalinism gave way to hope for a

---

[9] According to Molotov, his most loyal squire and Minister of Foreign Affairs, *"[Stalin thought that] World War I has wrested one country from capitalist slavery; World War II has created a socialist system; and the third will finish imperialism forever"*.

new dawn. In reality, little changed, for better or worse. Entrenched within its own impenetrable borders of men and ideas, the USSR remained under the total control of a single party and in search of a new single leader. It seemed that the Steel Man's dolphin might be Georgy Malenkov, but from the ranks of the CPSU emerged the somewhat awkward figure of Nikita Sergeyevich Khrushchev, already proconsul in Ukraine and party secretary, who would manage to concentrate all power in himself over the next five years of bitter internal struggle for succession.

A man of humble peasant origins, average stature, modest appearance, rough and resolute manners, and no education (he had attended only two years of elementary school), Khrushchev knew how to play his cards with great skill and the same disregard for danger that he had shown as a political commissar during the siege of Stalingrad. The year 1956 was decisive for his ascent to the "throne". Beginning with the bold denunciation of Stalin's crimes and aberrations in a closed speech on the cult of personality delivered on February 25 to the delegates of the XX CPSU Congress, it continued in the fall with the bloody Hungarian uprising and the concurrent Suez crisis. Actions and events that had weakened his position in the palace. He needed an ace in the hole to regain the consensus and grab the richest prize, the prime minister's chair. Almost unexpectedly, space came to his rescue.

For some time, Khrushchev had been nurturing the idea of curbing the massive spending on armaments that the escalation of the Cold War had imposed, and inaugurating a new policy of peaceful coexistence with the West. A position shared by American President, Dwight Eisenhower.[10] This

---

[10] Ike, the soldier president who, together with the Red Army, had reconquered Europe and liberated it from the Nazi-fascism, was tired of blood. In April of 1953, shortly after Stalin's death, between one round of golf and another he reflected on peace and the role that his great country, emerged victorious from the most devastating conflict of all time, had to play for the triumph of justice in a now bipolar world, where the other pole was the "empire of evil", to use the expression stolen by Ronald Reagan from *Star Wars*:

The Soviet government held a vastly different vision of the future. In the world of its design, security was to be found, not in mutual trust and mutual aid but in force: Huge armies, subversion, rule of neighbor nations. The goal was power superiority at all cost. Security was to be sought by denying it to all others. The result has been tragic for the world and, for the Soviet Union, it has also been ironic. The amassing of Soviet power alerted free nations to a new danger of aggression. It compelled them in self-defense to spend unprecedented money and energy for armaments. It forced them to develop weapons of war now capable of inflicting instant and terrible punishment upon any aggressor [...] Every gun that is made, every warship launched, every rocket fired signifies, in the final sense, a theft from those who hunger and are not fed, those who are cold and are not clothed. This world in arms is not spending money alone. It is spending the sweat of its laborers, the genius of its scientists, the hopes of its children [...] This is not a way of life at all, in any true sense. Under the cloud of threatening war, it is humanity hanging from a cross of iron.

objective, however, required a rebalancing of the forces in play, which were unbalanced by the apparent air supremacy of the Americans and their allies. In short, it was paradoxically necessary to invest in weapons in order to defuse the senseless spiral before it led to a general holocaust.

The shortcut could only be the intercontinental ballistic missile. Thus was born within the OKB-1, the Experiment and Design Bureau[11] of the NII-88 located in Podlipki (renamed Korolev in 1996 in honor of the chief designer), in the Moscow district, the project that would lead Korolev and his collaborators to the construction of the R-7 Semyorka, a powerful and long-lived vector. Even today, the Russians use a modified version of this "old faithful" for themselves and in international collaborations.

As the work progressed, Korolev began to entertain the idea that a multi-stage ICMB could also be used to place satellites in Earth orbit and, why not, to go to the Moon and nearby planets, Venus and Mars. In 1954, in a letter to the Soviet government, he wrote:

I would like to draw your attention to the memorandum by Comrade M.K. Tichonravov, entitled Memorandum on an Artificial Earth Satellite, and also to the material sent by the United States on the work in progress there in this field. The current development of a new product [the R-7] allows us to consider the possibility of developing an artificial satellite in the near future [...] It seems to me that at the moment there are conditions [...] for carrying out initial exploratory work on a satellite and more detailed work on complex problems related to this goal. We await your decision.

Here is the incipit of the Tichonravov *Memorandum*:

At the present time, there are realistic technical possibilities of achieving speeds sufficient to create an artificial Earth satellite with the help of the Article R [the R-7 ICBM]. The most realistic and achievable in the shortest time is the creation of an artificial satellite of the Earth in the form of an automated device, which would be equipped with scientific apparatus, have radio communication with the Earth, and orbit the Earth at a distance of about 170-1100 km from its surface. We will call such a device the simplest satellite [Sputnik in Russian]. The simplest satellite is imagined as an apparatus without people, moving in an elliptical orbit, and intended for scientific purposes. The weight of such a satellite could be of the order of two or three tons, taking into account the scientific apparatus. As will be seen below, the ways to realize the simplest satellite are basically clear at the present time. Undoubtedly, some issues will require further research, but in any case, it is possible to speak of the creation of a technical design of the simplest satellite. The

---

[11] Made independent from NII-88 in 1956.

timeframe for its realization depends only on the time for the creation of the article R, with the help of which it is possible to obtain the necessary speed. The design of the satellite can be carried out in parallel with the creation of such an article R. If work is carried out in this direction immediately, the creation of the simplest satellite can be carried out in the near future.

Prophetic words. At the same time, the Soviet Union's powerful Academy of Sciences had also begun to consider the peaceful use of future ICBMs as launchers for artificial satellites, with an eye toward science and related technologies and a wink at future military uses of space. As a result of these trends, parity in defensive and offensive systems did not end the race for supremacy by each of the two blocs. Instead, a new competition emerged for an intangible prize: prestige. For politicians in both the Kremlin and the White House, it represented an extraordinary added value. In search of more solid crutches, Khrushchev was indeed willing to play the *Sputnik* card, especially since the United States had agreed to launch one or more satellites on the occasion of the International Geophysical Year (IGY).

The beginning of a new deal for peaceful (so to speak) rocketry is conventionally placed in April 1950. James Van Allen, who was then working at the Applied Physics Laboratory at Johns Hopkins University, had gathered a number of colleagues at his home for a dinner party. One of the guests, American physicist Lloyd Berkner, proposed the idea of an international coordination to study the physical properties of the Earth and the planet's interactions with the Sun. A research topic that technicians now call space weather or space meteorology. The plan was to take advantage of the new technologies that the war had brought to science. It included the possibility of establishing a permanent laboratory in orbit around the Earth, as it would provide a panoramic view of the planet, free of atmospheric filters. Approved by the International Council of Scientific Unions, the request became plausible with the death of Stalin and would become the largest cooperative study of the Earth ever undertaken. At the suggestion of another Van Allen guest, the English geophysicist S. Chapman, it was decided to designate the period between July 1, 1957, and December 31, 1958, as the International Geophysical Year, because the activity of the Sun[12] would have been particularly intense during those 18 months.

On June 29, 1955, President Eisenhower declared that the United States would contribute by launching "*small Earth satellites*". The announcement

---

[12] Solar activity is associated with an 11-year cycle during which the star's magnetic field completely reverses. It manifests itself in changes in solar radiation, coronal mass ejections, and the size and number of sunspots.

was followed in August by a commitment from the Soviets to do the same. It came during a meeting with journalists at the Soviet Embassy in Copenhagen on the occasion of the Sixth Congress of the International Astronautical Federation, which had been founded in London four years earlier with the cheerful motto "*Astronautics for Peace and Human Progress*". Physicist Leonid Sedov, spokesman for the Soviet space program, announced his country's intention to put a satellite into orbit during the IGY. Of course, no one in the West paid much attention to it. The American dream was dazzling everyone, including the Americans, who were contemplating the monopoly of the future by a new civilization of stars and stripes.

Confident of their superiority, the Yankees rested on their laurels, allowing sterile disputes within the armed forces as to which specialty, naval or aeronautical, should engage in an enterprise that put prestige, money, and a lot of power on the table, and what the purpose of the enterprise should be, purely scientific or also military.

The beep of *Sputnik 1* woke everyone up. The Soviets had "*spit first*", as it was called in barracks jargon. The Reds' satellite had entered orbit and would remain in the sky until January 4, 1958, when the progressive erosion of its speed by the atmosphere would cause it to crash to Earth in flames, after 1440 orbits and a covered distance of 70 million km, with a route that even crossed the American sky. A home invasion confirmed with concern by specialists at the Naval Research Laboratory in Washington, who had immediately set to work on the telemetry data. This time the Russians had nothing to hide. On the contrary, they had made sure that everyone, really everyone, could see their triumph first hand. *Sputnik* had been polished to a mirror finish so that it could be seen after sunset or before dawn, like another star in the sky, pendant to the red star that shone on the highest spire of the Kremlin.

To beat the clock and keep the weight under control, Korolev had drastically reduced the onboard instrumentation of his *Prosteyshiy Sputnik 1*, or Simplified Fellow-traveler. When, together with Tikhonravov, he began to design a payload capable of meeting the geophysicists' expectations, he had to face the unreliability of the numerous subcontractors responsible for the various equipment. Time passed and there was a risk of being overtaken by the Americans, who, according to intelligence reports, were also working on a launch with their Vanguard rocket.

"*What if we made it a little lighter and a little simpler?*", asked Tikhonravov. "*Thirty kilos or so, or even lighter?*". The question, addressed to Korolev, triggered the spring. It was the Columbus egg from which a purely demonstrative satellite was born, supervised by Tikhonravov while Korolev managed the whole project. In addition to the radio, which worked for 57 days before the batteries ran out, *Sputnik* carried only a thermometer. Really very little to celebrate the IGY, but enough to make an important discovery that instead slipped through the hands of Soviet scientists. They had noticed a systematic pairing of periodic interruptions in radio transmissions with the altitude of the satellite, which varied greatly because of its elongated orbit, but they did not draw any conclusions. It would be Van Allen's merit to have intuited the existence of a large equatorial doughnut of charged particles around the Earth. Entering it at the apogee of the orbit, *Sputnik* was silenced by the interaction of its radio transmissions with the plasma of the famous belt that today bears the name of the American researcher, at that time particularly low due to the intense solar activity.

After the feat was accomplished, Tikhonravov noted in his diary: "*The newspapers are writing about the launch of Sputnik*". That's all. He was a man "*without haste, scrupulous in his judgments, [and] capable of reflection. He never imposed his ideas on anyone and never raised his voice*". Korolev, instead, had an impulsive and volatile character and was feared by all. He would get angry over the smallest nonsense, but never with his friend Tikhonravov. Together with Glushko, the two of them had performed the miracle.

The news of *Sputnik* reached von Braun while he was sipping a cocktail in the company of high-ranking Pentagon officials at the mess officials at Redstone Arsenal in Alabama, a sort of luxury *sharashka* for the German scientists of the Paperclip program. "*I will be damned!*" was his reaction when he found out he had missed the boat. Eisenhower, who had just been re-elected to the White House, tried to downplay the value of the feat in the press so as not to increase the growing fear of the Soviets. However, the "*useless hunk of iron*", as he called it, showed the scientific and technological strength of an enemy considered dangerous and treacherous, and gave reason to those who regretted the great inquisitor Joseph McCarthy and the Senator's hunt for communists conducted in the early 1950s.

After so many past troubles—the Berlin crisis, the fall of Chiang Kai-shek in China, the nuclear stalemate of the Soviets, and the bloody Korean War—there was reason to hope in the new leadership of the USSR for a softening of the Cold War. But now, the little *Sputnik* rekindled the reasons for the race. More than that, it was the sign of a rising red tide.

An affront and a danger that the White House tried to minimize: "*[We are not interested in participating] in an outer-space basketball game*", or *Sputnik* has not increased "*our apprehensions by one iota*". But anxiety and discontent had now festered in the souls of Americans, as witnessed by an open letter entitled *The Lessons of Defeat* and sent to the *New York Herald Tribune* by Bernard Baruch, a prominent figure in federal politics and on Wall Street:

While we devote our industrial and technological power to producing new model automobiles and more gadgets, the Soviet Union is conquering space [...] Suddenly, rudely, we are awakened to the fact that the Russians have outdistanced us in a race which we thought we were winning. It is Russia, not the United States, who has had the imagination to hitch its wagon to the stars and the skill to reach for the Moon and all but grasp it. America is worried. It should be.

Behind the Iron Curtain, the celebrations went wild, but the two main players, Korolev and Tichonravov, had to remain in the shadows, treated as military resources. From the general triumph they gained only a few more medals, useful for salary purposes. Sergei also received a rehabilitation that finally erased the stain of his arrest and long imprisonment.

Reserved and reluctant to exhibit, a true Stakhanovist and aware of the need to protect the secret formulas of his studies, shadow in the shadow, Tichonravov always stayed behind the shoulders of Korolev, the strong man for whom he always showed reverence and affection. The only note of emotion in his dry and cryptic work diary was when, in 1966, he had to report the untimely death of his great friend.

When the Nobel Committee wanted to reward the creators of *Sputnik*, Khrushchev refused to name his thoroughbreds and suggested instead that the coveted recognition be given to the entire Soviet Union. But this request was incompatible with the rules of the Nobel Foundation, which only honors living individuals, and above all it was unthinkable for the countries of the Western bloc. So no Nobel![13] This unpleasant affair greatly upset Korolev, who nevertheless continued to love his homeland first and foremost, and confined himself to saying, between the ironic and the bitter: "*When I bite the dust, they won't even write my obituary*". It would become a mantra, but

---

[13] During an interview Sergei Khrushchev said:

The Nobel Prize committee decided to give an award to Sputnik's 'chief designer', but first it needed the person's name. The committee requested it from the Soviet government. My father weighed his response carefully. The matter was complicated, and his concern wasn't confidentiality. The Council of Chief Designers was in charge of all space projects. Korolev was the head of the council, but the other Chief Designers – more than a dozen – considered themselves no less significant. My father understood that the Chief Designers were ambitious and jealous people. If the Nobel committee were to give the award only to Korolev, my father thought, the members would fly into a rage. They would refuse to work with Korolev. A well-organized team would collapse like a house of cards, and the hopes for future space research and missile design would be dashed, threatening the country's security. As my father saw it, you could order scientists and engineers to work together, but you couldn't force them to create something. In the end, my father told the Nobel committee that all of the Soviet people had distinguished themselves in the work on Sputnik, and that they all deserved the award. Korolev was offended but kept silent. The Nobel Prize went to somebody else.

fortunately he was wrong. In due course, he received solemn state funerals, and *Pravda* (Truth), the official newspaper of the Communist Party of the Soviet Union, devoted two full pages to him. Death had suddenly removed the state secret. Thus, for the first time, the Soviet people and the rest of the world knew the name and face of the man who flew the red flag with the hammer and sickle above all others.

Dog's Heart

The ray of the Moon, behold, comes to call me.
Edmund Rostand
A dog's spirit dies hard.
Mikhail Bulgakov

Moscow, on an unspecified day around New Year's Eve 1957. The city's dog catchers had once again been called out on a government job. They had to capture a few dozen small, healthy stray females. Without asking why—wise practice in those times and places to avoid unnecessary trouble—they diligently did their job, chasing the cold animals as they wandered among the snowdrifts in search of food, and delivered the prey to an official. Little did they know that among their prey, huddled in a corner of the large cage, was a little dog destined for eternal fame. She would be the first creature to enter Earth's orbit.[1]

For six years, the Soviets had been using dogs to test the reaction of living organisms to the harsh conditions of space: violent accelerations, lack of gravity, penetrating radiation, and, why not, even panic. The poor animals were launched on suborbital trajectories, parachuted back to Earth, recovered if all went well, and analyzed. A cruel activity, perhaps, but indispensable to overcome the skepticism of the great luminaries of medicine, convinced of the impossibility of human space flight.

---

[1] By convention, the boundary between the atmosphere and space is set at 100 km above sea level. This is called the Karman limit, from the surname of a Hungarian engineer-physicist who demonstrated that, at these altitudes, an aircraft would have to travel at a speed greater than orbital speed in order to have lift.

The test subjects were recruited from among the stray dogs, in the belief that the harsh living conditions on the streets of a big city would select the strongest. In short, if these strays could survive the Moscow winter, they might be able to make it in space. They had to be females, because they are usually more docile than males and easier to handle for their physical needs, small in size for weight and bulk reasons,[2] and with a colorful coat to look good in the newsreels. Legend has it that one of the dog catchers, perplexed by the long list of requirements, ironically asked if the dogs also had to be able to bark in C major. An unlikely story, as such a comment could earn him a one-way trip to Siberia. After all, Stalin did not live in vain.

The excitement caused by *Sputnik* had made Khrushchev realize the enormous propaganda value of the space, more effective than any other message in promoting the image of triumphant socialism. For this reason, the First Secretary had bypassed the chain of command and turned directly to his *glavny konstruktor* to ask for another miracle, just as grand and immediate. Korolev was looking forward to nothing else. He had a whole rosary of projects more or less ready for a long series of record-breaking space ventures. Khrushchev bought them all, starting with the one to put a dog into orbit.

But it had to be done quickly in order to properly celebrate the anniversary of the Bolshevik Revolution, which would fall on the coming November 6 (Julian calendar). The supreme leader was served on time. At dawn on November 3, 1957, before the Sun had risen over the Kremlin and just one month after the launch of the first artificial satellite, a capsule carrying *Kudryavka* (Little Curly), a small Muscovite dog, took to the sky. Westerners will rename her Laika,[3] having confused her name with that of the breed. A former tramp with a future as a space hero.

Laika was a three-year-old female weighing six kilograms. She was a cross between a sturdy Siberian husky, a sled dog, and a terrier, the classic companion of hunters. She had very sweet eyes and a gentle character, so much so that she never barked at other dogs. Along with *Belka* (Squirrel) and *Strelka* (Little Arrow), she was part of a group of three animals chosen as the "expendable" crew of the second orbital flight in history. A program that, because of its inevitable cruelty, lent itself to the instrumental criticism.

The protocol for training these quadrupeds for space was not much different from that used for human cosmonauts. Exposure to the accelerations of a centrifuge (a cell attached to a long rotating arm), deafening noise and very strong vibrations that the animal would have experienced at launch, and prolonged confinement in increasingly small spaces. And then medical visits,

---

[2] A requirement of jockeys, also applied to the selection of the first cosmonauts.

[3] Some say that Laika is also a proper noun, meaning "the one who barks".

training to eat a highly nutritious gel that would have been the only source of food and hydration during the flight, and surgical operations to place some sensors under the skin to monitor vital functions (breathing, heart rate, and blood pressure). But no lessons in Marxism-Leninism, as would have been the case for human cosmonauts, in order to make them worthy of their glorious achievements and, above all, to make them know how to behave in public.

Three days before the launch, Little Curly was dressed for the journey, wired up, and placed in the padded capsule at the top of the R-7 Semyorka rocket to acclimate to the new environment. When the door closed for the last time, leaving the cold winter outside the cramped compartment, a shiver of emotion ran through the staff who had grown to love Laika and knew what sad fate awaited her. The ticket was for a one-way trip. There had been no time to prepare the module for a controlled return to Earth.

The food and oxygen should last a few days. Then, according to Tass, the little dog was to be euthanized with a quick poison *"to prevent her from suffering a slow agony"*. But Laika's fate was different from the Soviets' propaganda lies. *Sputnik 2* reached its orbit without problems, an ellipse with an apogee of 1660 km, almost double that of *Sputnik 1*, and released its payload of a good half ton, six times larger than the first Simplified Satellite. Another spectacular technical success of Soviet rocketry, which demonstrated to the Americans the extreme danger of the enemy, which had been greatly underestimated until then. In the midst of the excitement, Laika survived for several hours in weightlessness and even ate some food. Then, as would be known only after the fall of the Berlin Wall, the frantic beating of her heart stopped forever.

After two or three orbits around the Earth, the little dog had died from the consequences of the enormous thermal shocks between the periods of exposure of the capsule to the scorching Sun and those of immersion in the icy cosmic night. Perhaps the thermal insulation, damaged at the time of detachment, had not functioned properly. For five and a half months, the capsule orbited in a gentle spiral around the Earth on lower and lower orbits before disintegrating upon contact with the dense layers of the atmosphere. A life like so many others sacrificed for the progress of science, but this time in front of the eyes of the world and not behind the closed doors of a laboratory where no one sees, no one checks, and no one criticizes.

The heroic enterprise unleashed a new wave of enthusiasm in the USSR and the galaxy of satellite nations and sympathizers. For a people who had recently suffered 23 million deaths in a devastating war won by fighting tooth and nail, the sacrifice of a dog was secondary to the positive values of the mission. Laika became a kind of pop icon, a symbol of a system that had been the first to successfully launch "*the attack on space*" and had lifted into the sky a people that had once been just peasants and workmen. Images of the little dog were reproduced everywhere, from matchboxes to chocolate wrappers. Such was the power of the message that in 1961, after his historic flight, Gagarin is said to have jokingly asked during a banquet in his honor: "*I am still unaware who I am – the first man or the last dog*".

The Western world, instead, reacted to *Sputnik 2* with outrage and concern. The killing of an innocent animal, it was said, had revealed the true face of a ruthless and bloodthirsty red regime. Who knows—many thought— maybe Senator Joseph McCarthy was right in his anti-communist crusade? A truly bizarre world, capable of crying desperately over the death of a dog and remaining silent with guilty consent in the face of massacres of human beings. With an irritating hypocrisy, Americans, though zealous followers of the Bible, seemed to have forgotten Christ's admonition: "*Why do you look at the speck of sawdust in your brother's eye and pay no attention to the plank in your own eye?*". While Laika was dying, racism and oil were claiming countless victims in every corner of the planet and in the silence of the consciences of the well-meaning.

The mourning for Laika, celebrated for decades by singers and artists, was mostly a pretext to turn the fear of the common people into indignation. Despite reassuring statements, the Eisenhower Administration began to feel uneasy. The *Sputnik* crisis affected not only national pride, but more importantly, the security of a country that had felt protected by two vast oceans. "*We now know beyond a doubt that the Russians have the ultimate weapon: a long-range missile capable of delivering atomic and hydrogen explosives across continents and oceans*", was the worried remark of Democratic Senator Richard Russell of Georgia, chairman of the Armed Services Committee. The alarm came from the highest rooms of the Pentagon and from scientists of the caliber of Edward Teller, the father of the hydrogen bomb. Ike, the hero of the war against the Nazis, was rapidly and unfairly losing points: "*A smiling incompetent* […] *a 'do-nothing', golf-playing president mismanaging events*". There was an urgent need to take cover, to reassure and protect Americans in the face of new dangers.

First of all, it was mandatory to ascertain the state of the art of rocketry in the country. On November 25, a Senate committee chaired by Majority Leader Lyndon B. Johnson began six weeks of hearings on the "*missile gap*" between the two superpowers. The incipit of the future U.S. president was a kind of call to arms:

A lost battle is not a defeat. It is, instead, a challenge, a call for Americans to respond with the best that is within them. There were no Republicans or

Democrats in this country the day after Pearl Harbor. There were no isolationists or internationalists. And, above all, there were no defeatists of any stripe.

But the result of the discussion was a seemingly change in the perception of the problem. "*These opportunities reinforce my conviction*", Eisenhower wrote the next day in a document issued by his Science Advisory Committee, "*that we and other nations have a great responsibility to promote the peaceful use of space and to utilize the new knowledge obtainable from space science and technology for the benefit of all mankind*".

More concretely, the Deputy Secretary of Defense, Donald Quarles, declared: "*The firing of the Sputnik had convinced me that we are apt to get support from the people of the United States for a stronger program than I thought I was going to get support for last spring*". The Soviets had thrown down the gauntlet twice, and Ike, with Anglo-Saxon phlegm, between one game of golf and another had decided to pick it up. May the best man win—he thought—believing that he knew exactly who this man was. But for the time being, as senator Johnson's entourage now well understood, the shrewd general was sorely mistaken: "*The simple fact is that we can no longer consider the Russians to be behind us in technology. It took them four years to catch up to our atomic bomb and nine months to catch up to our hydrogen bomb. Now we are trying to catch up to their satellite*".

The two blocs, consolidated in the aftermath of the world conflict, were maintained in a state of precarious equilibrium. The encirclement of the Western sector of Berlin marked the first formal act of the Cold War. In response, on April 4, 1949, the Western powers had forged a pact, the American-led North Atlantic Treaty Organization (NATO), which was followed six years later by the Warsaw Pact, signed by eight communist countries. An ideal setting for a tug-of-war in which each participant wants to win by preventing the other from losing his head and going to the knife. Neither the Russians nor the Americans wanted to start a nuclear brawl.

It was time for the United States to take appropriate countermeasures and respond to the Soviets using the same weapons. The one available at the time was the Vanguard TV3, a three-stage missile designed for scientific purposes by the U.S. Navy. An evolution of the Viking missile series, born in 1947 from a contract of the United States Naval Research Laboratory to the aircraft and spacecraft factory founded by aviation pioneer Glenn Martin. This was the launcher on which Eisenhower relied when, in 1955, he dared to announce the production of small unmanned Earth satellites as a contribution to the United States' participation in the IGY.

Actually, there was also another possibility, represented by the three-stage rocket, two of them solid fuel, developed at Redstone Arsenal by the U.S.

Army Ballistic Missile Agency under the direction of von Braun, who by now had made a career in his new country. His old homeland had been plundered of both material and immaterial goods by the victors and then divided up like his hunting grandfather had done with deer.

For the Peenemünders, the journey to American shores was not as smooth as they had hoped, given their value on the military market. On arrival in the United States, at the end of a pilgrimage made more difficult by a certain desire for revenge on the part of the victors, they were parked at Fort Bliss in El Paso County, Texas, in basic quarters and under the supervision of the military police. There they were joined by von Braun, who had been duly cleansed of his past "sins" to avoid criticism from the righteous. As the Emperor Vespasian liked to say, *pecunia non olet*, money does not stink, but don't let your constituents know it, especially if they're Puritans.

For the German prisoners, things had begun to look up when the Korean War broke out. They were sent *en masse* to the Redstone Arsenal base near Huntsville, Alabama—a village with "*two restaurants in all*" and a town square where middlemen met cotton farmers—to design ballistic missiles for military applications. Nothing strange. Their mentor, a former SS major and provider of the deadly V2 to the Führer, had never stopped thinking about war and had suggested, in addition to civilian and scientific applications, the use of the Moon as a base for missile batteries aimed at the Earth.[4]

It was precisely this explicit warlike purpose that condemned the Jupiter rocket, the latest offspring of German brains, and advised against its use as a possible carrier of an artificial satellite. The White House did not want to expose itself to the inevitable accusations of imperialism, nor did it want America to enter the cosmos thanks to a former Nazi. The fear of alienating public opinion had even led the administration to prevent a full test of the Jupiter A rocket. On September 20, 1956, when it was launched from Cape Canaveral in Florida, its last stage contained sand instead of fuel. Space had to be conquered by a true American!

---

[4] The idea was taken seriously with Project Horizon, proposed in 1959 by the U.S. Armed Forces to establish a lunar outpost:

> *The lunar outpost is required to develop and protect potential United States interests on the Moon; to develop techniques for lunar reconnaissance of Earth and space, for communications, and for operations on the surface of the Moon; to serve as a base for lunar exploration, for further space exploration, and for military operations on the Moon, if necessary; and to support scientific investigations on the Moon.*

But Jupiter C did its job so well[5] that it made Korolev, who had no shortage of information about the enemy's moves, think that the Yankees were about to launch a satellite and that the next flight on August 8, 1957, was just a failed test. A suspicion that gave him wings. Like Napoleon, he could not lose a battle at the risk of losing his crown.

In this confused scenario, the result of too many players on the scene and the rivalry between the various branches of the U.S. armed forces, a Vanguard TV3 was launched from Cape Canaveral on December 6, 1957, with a tiny satellite on top: an aluminum sphere, "*the grapefruit satellite*", as Khrushchev would scoff, weighing a modest 1.4 kg and equipped, like *Sputnik*, with a radio transmitter and a thermometer. On "go", the rocket rose about a meter and then fell back on itself, disintegrating and severely damaging the launch pad. Inexplicably, the satellite was ejected from its casing and fell to the ground intact, continuing to emit its beep like a wounded animal. A disaster that the American press, notorious for giving no one a break, pounced on, inventing derisive nicknames like *Flopnik* and *Kaputnik*. On Wall Street, the stock of Martin, the company that built the rocket, plummeted and was suspended for excessive trading. The Bolsheviks gloated, and the Yanks began to fear the worst.

To remedy the situation, the White House decided to let go of the moral issues and put the Führer's former engineer in charge, as he seemed to be the only one who understood rockets in America. Under Van Allen's scientific supervision, a second satellite was prepared in record time. Called Explorer 1, it weighed 14 kg, 8 of which were instruments. In addition to the usual things, it contained a microphone to report the impact of micrometeorites and a Geiger counter to detect cosmic rays. Placed on the Jupiter C rocket, duly modified by the Army Balistic Missile Agency of von Braun in synergy with the Jet Propulsion Laboratory (JPL) in Pasadena, California, it was successfully launched on January 31, 1958, into a very flattened elliptical orbit with an apogee at 2550 km.

The winning move was repeated on March 17 of the following year with the successful launch of Vanguard 1 and its tiny "grapefruit", after a second failure a month and a half earlier. The flight was only a demonstrative act,

---

[5] Alan Shepard would later write:

Eisenhower, his science adviser James Killian, and others in the White House didn't want to be reminded that the rocket was the same damn Jupiter-C that could have placed a satellite in orbit more than a year before Sputnik. The Army was told to keep that information quiet – in fact, to change the name of the rocket, and Jupiter-C became Juno 1.

without any particular scientific and/or technological significance, but it boasted two records: the duration (it is the oldest of the satellites still flying today) and the use of an experimental energy source, the photovoltaic panels, which were to become common in space. Nine days later, a third satellite, virtually identical to Explorer 1, joined the group. It would be remembered for its contribution to the discovery of the Van Allen belts.

The Americans breathed a sigh of relief. However, the persistent gap with the Soviets, who were capable of putting a handful of quintals of payload into orbit compared to the few kilos of the Star-Spangled missions, was not lost on anyone. National security and technological prestige remained at stake. There was an urgent need to put more money into the space budget and spend it better, taking the toys out of the hands of the military, among other things, to eliminate the fighting between the different roosters in the henhouse. Eisenhower therefore decided to create a new civilian government agency, the National Aeronautics and Space Administration (NASA), transferring to it the skills, personnel, infrastructure, and funding of the old National Advisory Committee for Aeronautics (NACA), founded in 1915.

NASA became operational on October 1, 1958, and subsequently took control of the Jet Propulsion Laboratory, which was dependent on Caltech, where it was born, and the Redstone Arsenal Agency, which on that occasion changed its name to George C. Marshall Space Flight Center (MSFC) and acquired a new prestigious director, Wernher von Braun. The former SS major had become a popular public figure thanks to his intense activity in promoting the great space adventure and his television appearances alongside Walt Disney, and now he wore the clothes of the savior of his new homeland. In fact, he was an American citizen, by law since 1955. In his second life, the shadows of the concentration camps and the black hats with the skull symbol no longer lurked, but there were still rockets.

If von Braun was awake, his rival Korolev was certainly not asleep. Like a racing cyclist who turns back and sees an opponent approaching in the distance, the chief designer was ready to accelerate his program. The masterful pedal stroke came promptly in the early hours of May 15, 1958, when the usual R-7, constantly improved to increase its performance, launched *Sputnik 3* into orbit from the cosmodrome at Baikonur. The payload consisted of a complete scientific laboratory to explore the highest layers of the atmosphere and to carry out observations and geophysical measurements.

No indulgence in spectacle this time. The satellite was a cone as tall as a one-story house and as wide at the base as a double bed, weighing almost three times as much as the Laika capsule, to be placed in an orbit with an apogee of 11,000 km and a life expectancy of almost two years. It was the

Soviet contribution to the IGY. Korolev had tried to launch this complete lab first, but, as we have said, fearing being overtaken, he had changed plans and entrusted the debut to a shiny, talkative, nearly empty sphere. Soon after, Khrushchev had intervened with his demand for a resounding success, and the chief designer had pleased him with Laika. Now it was time to serve science, as promised.

And once again, the *glavny konstruktor* was alone at the front of the race. In the now distant group of chasers, von Braun led the sprint of the stragglers. For the Americans, 1958 ended with two more Vanguard failures, a successful launch of Explorer 4, a satellite dedicated to the study of the Van Allen belts, and then seven consecutive flops. One of them, however, deserves to be remembered because it was the first of NASA's new leadership and marked the beginning of the race to the Moon. At 3:40 a.m. on October 11, a powerful multi-stage rocket that combined the experience of the Thor intermediate-range ballistic missiles with that of the Vanguards lifted off from Cape Canaveral. It carried the Pioneer 1 probe with the ambitious mission of placing it in lunar orbit. A previous attempt had broken up in a fireball after only 4 min of flight. This time it went better, but due to a launch vehicle malfunction, the celestial pioneer only traveled a third of the way before falling back to Earth.

In short, the Yankees were firing their machine guns, but with terrible aim. The Russians fired sparingly, much less, but it seemed as if they were hitting the target with each new shot. In reality, the impressive efficiency stemmed mainly from the Kremlin's habit of announcing only successes to the world and carefully hiding failures in silence or under a cloak of lies that the impermeability of the Iron Curtain made relatively easy to guard. A manipulation of information that is common practice for those who hold power with a strong hand and that had become a science with Joseph Goebbels. That is why for a long time nobody knew that the Baikonur magician had also tried to reach the Moon, failing three times. As usual, the order for this adventure came directly from Moscow. Khrushchev wanted to give America another knockout blow so that he could negotiate a truce in the Cold War from a position of strength. He demanded it within a year, without inquiring, as autocrats often do, whether the time frame was reasonable.

Disciplined, the chief designer obeyed. A version of the R-7 rocket was prepared, modified by the addition of a third stage to enable it to reach the speed of 11.2 km/s necessary to overcome the Earth's gravity.[6] For the "*escape*" stage, which had to be able to take off in a vacuum, Korolev decided

---

[6] If you threw a stone upward with an initial impulse, it would eventually stop its ascent and then fall back to Earth. But if you had the strength to give it an initial velocity of at least 11.2 km/s

not to use the new engine developed by Glushko's team, which provided formidable thrust but used a highly toxic synthetic fuel.

This choice may have also been influenced by the constant contrast between the two roosters of the space coop, dating back to Korolev's imprisonment in the *gulag* labor camp. The whole world is one country, but the USSR was even more so, because there, intellectuals, scientists, and technicians in particular were under the watchful eye of the *Komitet Gosudarstvennoy Bezopasnosti*, the Committee for State Security, or KGB. They could not make mistakes under penalty of being accused of sabotage. Nor could they allow anyone to stick his head out too far: almost an anticipation of the atmosphere that pervades the dystopian film *Rollerball* (1975).

Anyway, Korolev preferred to stick to the old road of liquid oxygen and kerosene engines, renewed by the intervention of a genius designer, Semyon Korsberg (1903–1965). The Luna 8K72, a rocket as tall as a 13-story building, failed the first test by self-destructing 90 s after launch due to a structural defect. The second launch, almost simultaneous with Pioneer 1, went the same way. The rocket flew for a couple of minutes and then broke in two. On the third attempt, it fell back to Earth after four minutes due to a failure in the second stage. But on the fourth, January 2, 1959, the 8K72 rose majestically into the sky, aiming for the Moon. Korolev wanted to call his creation *Mechta*, which means "dream". In fact, it was the mirage of all mankind that was about to come true. The mission was later renamed Luna 1 to give a chronological sense to the names of the now numerous missions to the Earth's satellite.

The declared goal, more modest than that of NASA, was to reach the Moon with a spacecraft loaded with instruments and make all kinds of measurements in an environment still unexplored. On board the sphere, pierced by radio antennas like a futuristic Saint Sebastian, there was everything: a magnetometer, a micrometeorite detector, the standard Geiger counter, and a scintillation counter to detect electrically charged particles, in addition to the usual equipment for transmitting the collected data and reporting the probe's position to the control center. With a speed nearly twice that of Pioneer 1, the Soviet rocket had enough power to lift a scientific payload of over one hundred kilograms.

On January 3, 1959, when it seemed all but accomplished, a beaming Khrushchev announced to the Supreme Soviet that the USSR had "*traced the path from the Earth to the Moon*". A laboratory with a hammer and sickle tattooed on its casing had overcome the Earth's gravitational pull and was

---

(high enough to circle the equator in one hour), then the stone would forever free itself from Earth's gravity and surrender to another master.

about to touch the Moon, where it would leave behind two metal pennants bearing the Soviet emblem and the inscription "USSR January 1959". It was a first for all mankind. Radio Moscow broadcast the news in fifteen different languages.

The stunned Americans tried to verify it, hoping for a Russian bluff, but the news was confirmed the next day, while newspapers around the world were already celebrating the feat with huge headlines. JPL technicians at the Goldstone Deep Space Communications Complex in California's Mojave Desert had managed to pick up *Mechta*'s signal, weak as befits someone now very far from home.

Fortunately for the Yankees, the last stage of the rocket had shut down a moment too late, causing the probe to veer off course and miss its target. After 36 h of travel, instead of hitting the lunar surface, the Dream had passed without ever coming closer to the target than an Earth radius. A failure? Only partially, because by saving itself from crashing, the recalcitrant kamikaze had automatically promoted itself to an "artificial planet", the first man-made object to enter solar orbit. No small achievement for the "*Soviet people, committed to the development of socialist society in the interests of all mankind*". Khrushchev failed to see this and, like the fox with the grapes, lied by declaring that everything had gone according to plan.

But that was not the mission's only record. In fact, Luna 1 made it possible to discover that the Moon, unlike the Earth, has no magnetic field. It also measured for the first time the wind of charged particles coming from the Sun and sampled the density of the upper layers of the Van Allen belts. The record for the distance of a radio transmission was broken. The spacecraft stopped communicating with its contacts in Baikonur when it was already half a million kilometers from Earth.

NASA responded promptly with a flurry of launches. Ten attempts in all, one for each month beginning in February, half of which failed for one reason or another. As consolation prizes for a nation accustomed to winning, there was the first photo of the Earth from space, taken by a satellite launched by a Vanguard, and the first approach to the Moon by the Pioneer 4 probe on March 3, thanks to a German-speaking variant of the Jupiter rocket. It missed the target spectacularly, ten times worse than Luna 1, but this probe also became a satellite of the Sun. An opening to that deep sky of which the Americans would become the masters in the last quarter of the century.

Whimsical Selene remained untouched, but not for long. Nine months after the failed attempt, Korolev tried again, with virtually no changes to either the vector and the probe. There were some problems with the launch, which had to be postponed for three days. Then, on the night of September

13, 1959, the 8K72 took to the skies. After the usual 36 h of travel, during which scientific measurements were made, the probe crashed on the Moon, West of the *Mare Tranquillitatis*, one of the large basaltic formations caused by the impact of giant meteorites in the infancy of the celestial body.

When the radio transmissions suddenly stopped, the engineers at Baikonur understood that they had hit the bull's eye. A wave of unstoppable joy reached the farthest corners of the huge base. They drank and sang with joy. It was the first time a human artifact had touched another celestial body. Half an hour after the first impact, the third stage of the rocket, which had accompanied the probe at a distance, also fell on the Moon. Two birds killed with one stone, you might say.

As a pure theatrical coup, a sphere segmented into pentagons, similar to a soccer ball, had been placed on board. Each petal carried the same message that Luna 1 had failed to deliver. At the moment of impact, a small explosive charge was supposed to fragment the structure and project the plates all around, almost symbolically taking possession of the celestial object in the name of the people of the Union of Soviet Socialist Republics; as Columbus did when, bowing on his knees, he planted the flag of the Catholic Kings of Spain on the beach of the island of San Salvador in the West Indies. We do not know if this provocative toy worked, but perhaps one day, when humanity has progressed further, these little shards of memory will become a hunting ground for astral archeologists and collectors.

The uniqueness of the event overshadowed for a while an important discovery of that trip: the confirmation that the radiation present in space was not of such intensity as to endanger the health of future cosmonauts. This was what the doctors wanted to hear in order to give their blessing to a manned flight program.

The Yankees were not prepared to respond adequately. As per the script, they launched another satellite with the usual Vanguard, focusing on instrumentation technology and the use of solar panels to power it. The U.S. industry was unbeatable in this field. As soon as the Soviets copied this solution to produce energy, the Yankees would switch to fuel cells and maintain their leadership in this sector.

Less than a month after the success of Luna 2, Korolev was ready to deliver a real uppercut. Shortly before one o'clock at night on October 4, 1959, Greenwich time, the Luna 3 probe, an object the size of a refrigerator and weighing almost 300 kg, lifted off from Baikonur. Tass made an announcement, obviously posthumous, but strangely devoid of the usual propagandistic frills. The mission seemed too complex to allow the bear's skin to be sold in advance. There was a real danger of failure. Too many things could go wrong. Beginning with the scientific experiments. In addition to the now standard equipment, a camera with two lenses of 200 and 500 mm focal length was mounted outside the spacecraft. The pictures, taken on black-and-white film, would have been automatically developed on site and then scanned so that they could be sent by radio to a ground station.

The destination was the hidden face of the Moon, the one that no human eye had ever been able to see, because our satellite, trapped in what astronomers call a spin–orbit resonance,[7] rotates on itself in exactly the same time it orbits the Earth. Just as this book would if you held it up to your face as you rotate regularly about yourself. You would always see the cover and the back would remain hidden. This happens because with each of your full rotations, which is also one revolution of the book, it completes a full rotation on itself. The solar panels would have provided the necessary energy to power both the device and the radio.

The launch parameters were calculated so that the spacecraft could reach the Moon, turn around it and, thanks to a gravitational assist—a kind of kick given at the expense of the Moon's gravity—be placed in a geocentric elliptical orbit with a perigee of 500 km. In this way, the vehicle would have come near the Earth several times, enough to allow the transmission of images collected during the close passage of the lunar surface.

Almost everything went smoothly. On October 7, the spacecraft flew over the dark side of the Earth's satellite for 40 min and collected 27 images from a distance of 60,000 km, mapping about 70% of the hidden hemisphere. Then it set off to rendezvous with the big ears on Earth: radio antennas eagerly awaiting the safari report. But when it reached perigee, Luna 3 almost lost its voice. Its signal had become too weak to receive the scans of the precious documents.

When he found out, Korolev was in Tyuratam. He became furious. Despite the adverse weather conditions, which didn't allow safe flying, he immediately ordered to be transported by military aviation to the Crimea. There was the big antenna, which was supposed to capture the voice of Luna 3. The chief designer pulled out his fingernails. He wanted to retrieve the pictures at any cost. After fifteen days, the spacecraft appeared again for the regular appointment. Korolev's angry scolding of his technicians had an effect, because this time it was possible to download 17 low-resolution images. Better than nothing. Ten more days passed, and contact with the probe was completely

---

[7] Resonance occurs when the ratio between the orbital periods of two bodies (orbit-orbit resonance) or between the rotation and revolution periods of the same body (spin–orbit resonance) is a fraction of small integers. This ensures that the system is always in the same configuration at regular time intervals, commensurable with the periods involved.

broken, but it continued to make its rounds until it crashed to Earth in April 1960. Amen!

Although of very poor quality, the images[8] surprised the scientists. The far side of the Moon appeared to be heavily pockmarked by a kind of juvenile acne, but almost devoid of the dark spots called *Maria* (seas) that are characteristic of the visible face. Apart from that, they showed the same desolation. Nothing to do with the description given by the passengers of the spaceship in Jules Verne's *Around the Moon*: no forest or lake, as they seemed to see in a flash of light caused by the explosion of a passing asteroid. The mountain ranges, valleys, craters, and all the other features of the ten images lent themselves to a quick baptism by the Russians, who did not miss the opportunity for a move with a strong propaganda content. Thus, one of the seas was named Moscoviense, and some craters gradually took the names of famous cosmonauts: Gagarin, Titov, Tereshkova, and Komarov, and also of the noble father Tsiolkovsky. In this toponymic rapture, the Russians made a mistake. They called the Soviet Mountains a chain of heights that turned out to be non-existing. The name fell long before the Berlin Wall. It goes without saying that the operation was soon contested by the International Astronomical Union,[9] whose majority was pro-American.

It was enough to make old Ike eat his hat. The president had already endured Khrushchev's irony on the occasion of the Russian premier's last visit to America, which began on September 15, 1959, after the triumph of Luna 2. At the first meeting of the two leaders at the airport, however, Khrushchev had extended his hand to his rival: "*We have come to you with an open heart and with good intentions. The Soviet people want to live in peace and friendship with the American people*". He really needed peace to be able to carry out the reforms, especially in agriculture, that the country was waiting for. Among other things, the long relationship with the People's Republic of China, always an ally, was about to break off. The victories in the space race served to sell a less gloomy and more reassuring image of the USSR.

The outcome of the official meeting with Eisenhower at the president's summer residence at Camp David, Maryland, was formally more than reassuring. The joint statement issued at the end of the summit said that "*the exchanges of opinions would have contributed to a better understanding of the motives and position of each and thus to the achievement of a just and lasting peace*", and that "*the question of general disarmament is the most important*

---

[8] The mosaics were published in a *Lunar Atlas* inserted, in the late 1960s, in the *Great Soviet Encyclopedia* (BSE – *Bol'šaja Sovjetskaja Enciklopedija*).

[9] It is the international organization of professional astronomers recognized as the authority for assigning names to celestial bodies, including the features present on the surface of some of them.

one facing the world today. Both governments will make every effort to achieve a constructive solution of this problem". Fine words. But neither of the two leaders was really sincere. Each thought he could continue to play his own cards, Nikita in space and Ike in the much slipperier field of aerial espionage.

The Americans got a breath of fresh air with the March 1960 launch of another Pioneer, the fifth in the series. The original goal, a flyby of Venus, was ambitious enough to overshadow the last Soviet Moon mission. But a prolonged technical problem caused the launch window to be missed. In fact, when it comes to planets, the shot must be taken during the period when they are about to reach inferior conjunction (the moment when their distance from the Earth is at its minimum[10]). Having missed the fast train, they had to settle for a slow one, justifying the mission with an exploration of space between Earth and Venus that confirmed the existence of an interplanetary magnetic field. An information of considerable scientific value, but of little propagandistic weight.

Korolev, for his part, already had another full drum of bullets in his revolver. On May 15, 1960 at midnight Greenwich time, a real spaceship took off from Baikonur. So much so that it was named *Korabl-Sputnik 1*, or Vessel Satellite. For convenience, the Westerners preferred to keep the chronological series and simply call the mission *Sputnik 4*. It was an attempt to put into orbit and then recover a 4.5-ton module called *Vostok* (East), which was conceived to carry a human passenger on a round-trip in space. The module, equipped with its own engine, was placed atop an R-7, of which it was effectively the third stage. The flight test went well, but a re-entry failure pushed *Vostok* into a higher orbit, where it remained for about two and a half years before decaying and disintegrating. Nevertheless, it was the first step toward the human conquest of space. A stage in the race to the Moon that the Soviets would have won by a wide margin, wearing the Yellow Jersey, to use a *Tour de France* analogy, for almost the entire length of the competition.

The partial success prompted the chief designer to try a flight with animal and plant guinea pigs. The first attempt, made on July 28, ended in tragedy just seconds after liftoff when one of the rocket's side stages failed. Although the control room technicians had activated the rescue system for the guinea pigs in time, it did not work properly. Given the low altitude, the re-entry capsule's parachute deployed only partially. A lesson that convinced the designers to equip *Vostok* with ejection seats.

The second test, called *Korabl-Sputnik 2* or *Sputnik 5*, carried two dogs, *Belka* and *Strelka*, which we met while being trained with Laika, constantly

---

[10] Actually, a little before the minimum, since the flight time of a probe must also be taken into account; just as the duck hunter does when he shoots in front of the bird's beak.

observed by a camera in the control room, 40 mice, 2 rats, and a selection of plants. It was launched on August 19, 1960, at 5:38 a.m. from the pad that would be later named after Gagarin. The *Vostok* was placed in a slightly elliptical orbit at an average altitude of 300 km and was brought back to Earth after 21 h, using atmospheric friction as a brake and then a parachute. The animals were recovered in good condition.

Since the dogs showed signs of seizures during the fourth orbit, the doctors, underestimating the incredible adaptability of the human body, concluded that the first manned flights should not last longer than 3 orbits. However, it was a temporary malaise with no consequences for the animals. So much so that the following year Strelka gave birth to a litter of beautiful puppies. The mating was intended as another check on the effects of the flight. With the first female cosmonaut, Valentina Tereshkova, the need for a "mother for science" will arise again, as we shall see later.

A curiosity. One of the puppies was presented to Jacqueline Kennedy. This seemingly innocuous gift served as a tangible and poisonous reminder of Soviet superiority in space. Some even suggested that the pet had been implanted with microspies to steal the secrets of the White House. Such was the climate of the Cold War. After their deaths, the two cosmonaut dogs were embalmed to be displayed in a Moscow museum dedicated to space, following a practice that began with Lenin and continued with Stalin.

Khrushchev now loved his *glavny konstruktor* and protected him from the wrath of the Red Army bigwigs, who saw almost 50% of their budget wasted on beautiful but expensive machines[11] without even being able to count on an effective ICBM. The R-7, in its various modifications, remained a tool for science and research, but not for waging war. It took too long to arm (30 h to refuel) and depended too much on imposing launch sites that could be easily identified by the enemy. Korolev knew this, but showed no signs of concern.

This prompted Khrushchev to launch a new series of rockets that could compete with the Titans, the intercontinental ballistic missiles of the U.S. Air Force. They had to be more powerful than the R-12 and R-13 missiles designed for nuclear submarines, reliable and, above all, recoverable in well-protected silos and ready for launch within a few minutes. The task of designing them was assigned to Mikhail Kuzmich Yangel (1911–1971), a Ukrainian engineer under Korolev who had long hoped to outdo his boss and take his place. Supervision was entrusted to Marshal Nedelin, who had made the mistake of betting on the R-7 and now had to make amends.

---

[11] To get an idea of the investment required for an R-7, all included, consider that until recently it cost $50 million to buy one of three seats on a Soyuz for a few days' flight to the International Space Station.

Within a year, the two-stage R-16 rocket, thirty meters high, was ready for testing on a new launch pad added to the one where Korolev tested his missiles. It worked with a toxic and corrosive propellant that Glushko liked, which had the advantage of remaining liquid at room temperature. Just what was needed to have the weapon ready on demand. It could have been launched on American cities in a matter of minutes with the deadly yield of a 3-ton thermonuclear bomb. Bingo! Nedelin seemed eager to show it in action, to redeem himself in the eyes of the Party. He therefore decided that the test would take place in conjunction with the anniversary of the revolution, on November 7. To save time, many safety protocols were simply skipped and reports of criticality were ignored. A tragic mistake that would not remain unique.

At the moment of liftoff, on October 24, 1960, the second stage caught fire, causing the first stage to explode as well. Nedelin, who was nearby, was incinerated by a fireball. A total of one hundred people died and the new launch pad was badly damaged. Yangel miraculously survived because he had returned to the bunker to smoke a cigarette, which was strictly forbidden on the platform. One of the few cases where smoking, instead of killing you, saves your life.

Despite the disaster, the project was not halted. There was a quick investigation behind closed doors to determine causes and responsibilities, and one year later the R-16 made its successful maiden flight. The so-called Nedelin disaster, carefully hidden from the world,[12] had two major consequences. It freed Korolev from a dangerous competitor and convinced Khrushchev that to keep pace with America he would have to deploy a battery of medium-range Soviet ballistic missiles on the island of Cuba, a stone's throw from American soil. A decision that led to the famous international crisis and the long-distance duel with the new occupant of the White House, John Fitzgerald Kennedy.

The global scenario was changing, along with some of the main players. On November 4, 1958, the Patriarch of Venice, Angelo Giuseppe Roncalli, had been consecrated Pope John XXIII, and on November 8, 1960, the Catholic Kennedy had narrowly won the presidential election in the United States, also capitalizing on the American fear of Russian superiority in space control. The Democratic senator from Massachusetts had cleverly placed the blame on the Republican administration, accusing it of guilty delays and waste.

---

[12] It was said that Nedelin died in a plane crash, and so did the other technicians and workers.

The young president, a perfect embodiment of the American dream, inherited a new state of high tension between the U.S. and the USSR created by the downing of the U2 flying slowly at high altitude over Soviet territory on May 1, 1960. A sordid affair that had cost Washington much embarrassment and the collapse of the Geneva negotiations on controlled disarmament. Khrushchev had shown himself to be truly furious at the disloyalty of his adversary.

The year 1960 ended with another surprise move by Korolev that kept NASA officials awake at night: the launch of two probes in rapid succession toward the planet Mars. The goal was to study interplanetary space in the region beyond Earth's orbit and then, once on the Red Planet, to collect data and images. The delusions of Percival Lowell, the wealthy American who, at the beginning of the century, had preached the existence of an intelligent civilization on Mars, capable of exploiting the frozen waters of the poles with admirable networks of canals, would have received a decisive verification on the spot. But this time, luck did not smile on the chief designer.

Called Mars 1M by the Soviets and *Marsnik* (a combination of Mars and *Sputnik*) by the West, the missions were designed to place their respective probes in an initial parking orbit around Earth. From there, a fourth stage would push them into a heliocentric orbit with an apogee at the distance of Mars from the Sun. Both failed due to a malfunction of the *Molniya* (Lightning) rocket, a new four-stage version of the R-7 with a height of 43 m: a superfetation of the highly reliable original two-stage rocket which had been required by the need for greater overall performance. Tsiolkovsky's rocket train had more and more cars. Perhaps too many now.

Around New Year's Eve 1961, the world was once again on the brink of total war. But Korolev, undaunted, continued his increasingly ambitious program with the courage of a man who had faced death in Kolyma. Nothing could frighten him, and no one, not even his new family, could distract him from his work. In 1949, after Ksenia Vicentini had left him because he was unfaithful and too committed to work, he had remarried to Nina Ivanovna, a woman much younger than him who translated from English his scientific articles. Delightful creature whose love, however, was not enough to compensate for this giant's restlessness.

What was designed for Mars, he thought, might be good for Venus. The mysterious planet, shrouded in a dense mantle of clouds impenetrable to light, is even closer than Mars, although it takes about the same amount of energy to get there.[13] So on February 4, a *Molniya* rocket was launched with

---

[13] In general, it is cheaper to fly away from the Sun than to fly toward it. This may seem paradoxical, since the Sun is the body that exerts the much stronger gravitational pull in the Solar System. The

the intention of reaching Venus. Its probe was so massive that Westerners feared the Soviets were even experimenting with manned flight. The strategy was the same as with the *Marsniks*, and the result was equally disappointing, this time due to the fourth stage of the rocket. So the *Tyazheliy-Sputnik 4* (Heavy Satellite), or *Sputnik 7* as it was called in the West, remained parked in an Earth orbit.

The following *Sputnik 8*, also known as *Venera 1*, was launched eight days later, on February 12, 1961. This time everything seemed to go well. The spacecraft entered a heliocentric orbit, passing within four Earth radii of the blue planet. At least that is what is believed based on extrapolations of orbital data, because just one week after the launch, Earth stations lost contact with the spacecraft. The information possibly collected by the rich array of instruments on board this complex spacecraft could not be retrieved. A fate that would be shared by many of the subsequent Soviet space missions.

Two more launches of the *Vostok* spacecraft followed in March, classified as *Sputnik 9* and *10* (*Korabl Sputnik 4* and *5*). Both carried a dog and a mannequin so realistically made that it was labeled "dummy" to avoid being mistaken for with a dead astronaut. It was jokingly named Ivan Ivanovich.[14] Everything worked like a charm: even the re-entry. The mannequin's parachute opened immediately after the ejection seat was released, while the dog landed inside the capsule, also suspended by a parachute. This was the last test before the big gamble.

And the United States? In 1960, only pragmatically useful launches—a weather satellite, a passive one for intercontinental communications, Echo 1A, and, to appease the military, a spy satellite—and two failed attempts to orbit the Moon with the Pioneer P20 and P31 probes. But in January 1961, NASA responded to the Soviets' repeated winning serves with a backhand down the line that earned them a clean point. A Mercury-Redstone rocket, similar to the one successfully tested 45 days earlier, lifted off from Cape Canaveral for a 16-min suborbital flight carrying Ham, a Cameroonian chimpanzee selected from a group of primates trained by the U.S. space agency to serve as guinea pigs. Despite a series of glitches and malfunctions, the launch was a success. The first in a series of small steps to catch up with the Russians.

Lady Luck was evening the score with the Yankees, after being shamelessly on the Soviet side. After reaching apogee, the capsule turned to face re-entry,

---

fact is that when spacecraft leave the Earth, they share the speed at which the planet orbits the star, which is 30 km/s. This velocity must be compensated for in order to fall freely toward the Sun.

[14] Slang name used to indicate an individual with an unknown or secret identity, analogous to John Doe in the United States and NN (Latin "*nomen nescio*", I do not know the name) in Italy.

slowed by a large parachute in the last five minutes of the dive into the atmosphere. When a U.S. Navy helicopter picked it up, barely afloat in the waters of the Atlantic Ocean and in danger of sinking due to a leak, it was a relief to find that Ham had passed the test with flying colors.

In the photos taken at the opening of the spacecraft and distributed to the media, the poor animal, which had been exposed to weightlessness and tremendous accelerations, even greater than those for which it had been trained, showed a bewildered face. But it was alive and hungry! A step forward for the Yankees in the thrilling match with the Soviets. But once again, the game went to the players from behind the Iron Curtain, who were on the verge of winning the first set.

The Red Icarus

I am Gagarin. The first to fly, and you flew after me. I have been given forever to sky and Earth as son of humankind.
Evgeny Yevtushenko
Flectere si nequeo Superos, Acheronta movebo. (If I cannot bend the will of Heaven, I shall raise Hell).
Publius Virgilius Maro

Here we are again at Baikonur. It was just dawning, that Wednesday morning of April 12, 1961. It was just dawn in Tyuratam. The sky over the steppe, almost completely clear of clouds, promised a beautiful day, and the fresh air stimulated the appetite. Korolev had not slept all night, tossing and turning in bed. A passionate and fiery man, he could not put his worries aside for a moment.

Instead, Yuri Gagarin (1934–1968), he who would later be called the Christopher Columbus of the skies, and his backup Gherman Titov (1935–2000), were still sleeping soundly in their reserved room when they were awakened, at 5:00 a.m., to begin preparations for one of the most daring feats of all time. The night before, they had played billiards for a while and then dined in the company of their doctor, squirting into their mouths the "*cosmic food*" contained in some toothpaste-like tubes. Korolev had come to greet them, and to ease the tension, more his than the two carefree young men's, he had joked: "*In five years there may be a trade-union pass needed to fly into outer space*". "*There was no trace of alarm in him*", would say later Gagarin. "*He was confident of me, as I trusted him*". Hard to believe!

M. Capaccioli, *Red Moon*, Springer Praxis Books, https://doi.org/10.1007/978-3-031-54760-7_7

At bedtime the two cosmonauts were offered a sleeping pill: "*Guys, do you need help falling asleep?*", the doctor had asked them, but both had refused. The long, peaceful sleep, "*without nightmares and without dreams*", confirmed the deep inner serenity that the chief designer so much wanted in his colts.

Two nights earlier Gagarin had written a letter to his wife Valentina and his two daughters, Lenochka and Galochka, "*to share the joy and happiness*" he felt. "*Today a governmental commission decided to send me first to space. [...] I'm so happy; I want you to be happy with me. A simple man has been trusted such a big national task—to blaze the trail into space! Is there anything bigger to wish for? This is history, a new age*". He also advised Valechka that if the flight failed, she should raise their daughters "*not as some lazy mommy's girls, but real persons who can handle anything life throws at them. Make them worthy of the new society—communism. The state will help you do it*". With this request to his wife, Yuri wanted to confirm his faith in a social system capable of transforming a peasant into an eagle. But he was not afraid of dying; on the contrary, he was very optimistic: "*I trust the hardware completely. It will not fail*". The letter, which was to be delivered only in case of an accident, was opened years later, after Gagarin's unexpected tragic death. He had closed it with a prophetic "*I hope you never see it*".

After a breakfast of meat paste, blackcurrant jelly, and coffee, the two young men were taken to the medical center for a final checkup, which they both passed without problems. Most likely, Titov had secretly hoped that a last-minute minor illness of his comrade would allow him to snatch the role of lead pilot. At stake was the chance to become the first cosmonaut in history, an extraordinary opportunity he had missed at the very end of the long preparation phase. "*It is hard to decide who is to be ordered to face almost imminent death*", General Kamanin had once said, "*but it is as hard to decide who of the two or three equally worthy candidates is to be made a world celebrity, destined to go down forever in the history of humanity*".

Gherman always trusted that he would be chosen first because of his better qualifications. Instead, his comrade Yuri, now within sight of the finish line, had won, presumably because his humble origins played into Khrushchev's hands. Should the venture succeed, the leader would be able to boast of the merits of a socialism capable of advancing anyone, regardless of back-ground. The ranking imposed by the Kremlin was also shared by the *glavny konstruktor*, for a very different reason. Gherman was stronger than Yuri and could be more usefully employed in a future and more demanding mission.

But how were these young heroes selected? Through an operation that had begun two years earlier at the Soviet Air Force Research Institute (NII VVS—*Nauchno-Issledovatelsky Institut Voyenno-Vozdooshnykh Seel*) in Moscow. A committee of experts had defined the characteristics of the candidates. They had to be military pilots with extensive flying experience. This condition, which the Americans also imposed for their recruitment program, was based on the reasonable belief that the ability to tame a supersonic aircraft in any weather and light conditions and to handle emergencies with coolness and speed was already an effective filter for selecting the strongest. They also had to be of contained size and weight (they should not exceed 170 cm in height and 70–72 kg in body mass), under 30 years of age, in perfect health, even without surgical scars, and—get this—members of the Soviet Communist Party.

Overall a good set of rules, judging by the fact that of the first 20 candidates selected, 11 would actually fly. Unfortunately, three would die in various accidents. Apparently, one of the criteria not taken into account for lack of objective a priori indicators was one of the qualities Napoleon valued most in his generals, luck. In 1962, five women joined the group, including Valentina Tereshkova.

Candidates, usually recommended by an immediate superior or Party official, were required to submit their family's medical records. If deemed suitable, they were subjected to rigorous medical examinations and demanding psychological tests under extreme stress conditions. Long sessions in the centrifuge and on vibrating platforms amid deafening noise to simulate launch and re-entry conditions, exposure in low-pressure chambers, and decompression tolerance tests. This torture was necessary to ensure their survival in real situations and thus the success of the mission. About half of the candidates were discarded in the initial phase, more than half of the rest after the medical examination, and another 10% during the first month of training, mostly for failing the stress tests.

The selection criteria are well summarized by Korolev's own words.

It is not romantics for the sake of romantics that should provide the groundwork for a final decision when the selection of the first Soviet cosmonauts is in progress. Space rejects such people. Patriotism, bravery, modesty, ability to instantly make sound decisions, willpower of iron, knowledge and love towards other people – these should be the basic traits of character.

The few survivors were subjected to a period of hard training at a primitive center, which was replaced in July 1960 by the well-equipped Star City, a closed military townlet located in the Moscow area near a military airport and a major railroad. The facility and the program were placed under the control of Nikolai Kamanin (1908–1982), legendary pilot, polar explorer, and hero of the Soviet Union, who held this position until 1971.

A tenacious man and a convinced communist, for twelve long years Kamanin would have spent his inexhaustible energies fighting the indifference of the high aviation commands to space activities and the preference of Korolev and his engineers for automatic guidance systems that limited human intervention in the control of spacecraft. His *Diaries* remain an important source of information, albeit biased, about events that were obscured or distorted by the heavy filter of Soviet propaganda. They tell of heroic deeds, epic struggles in the myriad committees of the USSR's bureaucratic galaxy, sleepless nights before launches, big parties after triumphs, and carefree excursions in moments of freedom in the steppe around Baikonur.

The training program was designed to improve physical fitness, coordination, and adaptation to weightlessness, as well as the ability to withstand various environmental challenges such as high acceleration, high temperatures, and oxygen deprivation. At the first sign of physical or psychological weakness, the candidate was mercilessly eliminated. There were also psychological tests for emotional stability (for instance, two weeks of environmental isolation), suggestibility, working memory on sequential tasks under external interference, and the ability to respond to stressful conditions by maintaining the ability to distinguish a signal from background noise even under time pressure.

Gagarin and Titov had passed all the tests and were first in a group of six. It was up to one of them to board the *Vostok* for a short orbital flight, the first in history. A unique chance to become part of the legend. The decision was made just two days before the flight. "*It's Yuri's turn*", Korolev said on April 9, during a briefing of the cosmonaut team. "*Gherman will be the reserve and will fly on the next mission*". No one dared to comment, but Titov was very upset. You can even see it in the scowl on his face in the historical footage of the event.

"*I was frustrated, of course, because up to the last minute I thought my chances were high enough that I could have been the commander of the Vostok capsule*", he would confess later, still prey to bitter professional jealousy. Perhaps Gagarin's words still rang in his ears, like salt on a wound:

Am I happy to be starting on a space flight? Of course I am. In all times and all eras man's greatest joy has been to take part in new discoveries. I would like to dedicate this first space flight to the people of communism, a society which our Soviet people are already entering, and which, I am confident, all men on Earth will enter.

Surely Gherman would have wanted to say those glorious words himself.

After the medical checks, the two cosmonauts were taken to a room where they were dressed in their flight suits. It was a relatively complex and ritualistic operation—almost like putting on the matador's robe before a bullfight—that began with attaching some electrodes to the body to monitor vital functions. Then, the two cosmonauts received white silk underwear, a thermal undersuit, "*blue, warm, soft, and light*", and finally the pressurized space suit in bright orange, designed to make them highly visible in case of an emergency. The boots were fastened, a special soundproof helmet was put on, and a white helmet with a wide, transparent visor that could be lifted was placed on top. To complete the outfit and seal the spacesuit, only the gloves were missing. It was decided not to let the cosmonauts wear them for the time being, in order to have their hands free before the flight. The two also received a pistol each, to be used in case the descent into a hostile area required it.

The spacesuit was airtight. A precaution demanded by the doctors to minimize the risks in case of decompression, not shared by Korolev for reasons of weight and because he feared that the realization of the garment could delay the launch. "*Khorosho [ok], go ahead and do it*", he finally decided, "*but be on time! I will not wait*". They did it on time, because everyone knew that Sergei Pavlovich never spoke in vain. The white helmet bore four capital letters: CCCP (USSR in Cyrillic letters). It is said that they were painted at the last moment by an engineer with particularly good handwriting, because someone had pointed out that in case of landing outside the territory of the USSR the cosmonaut could be mistaken for a spy. Everyone had in mind the recent case of Lieutenant Powers, the American pilot who was captured after shooting down his U-2.

The chief designer also appeared in the room, "*tired and worried because of the sleepless night*", Gagarin later recalled. "*But from time to time a benevolent smile would light up his tense face. I wanted to embrace him as if he were my father*". While he was busy with the dressing, Yuri got the last advice from an experienced parachutist who had tested the descent model based on the ejection of the cosmonaut from the capsule. The final decision to use this procedure was made in the last days. The initial hypothesis was to land the passenger inside the spacecraft, but Korolev doubted that a man could survive the impact with the ground. This change in procedure created a problem. The rules contained in the special international agreement guaranteed by the *Fédération Aéronautique Internationale* (FAI) stated that in order to apply for an aeronautical record, pilots had to remain in their aircraft for the entire duration of the flight. "*No matter*", Korolev thought, "*if necessary, we will gloss over this issue*". And so they did.[1]

---

[1] The Soviets kept quiet about the landing methods and the FAI accepted Gagarin's record. But with the following flight of Titov the problem came to light. After a long investigation and futile lobbying by the Americans, the rules were changed to preserve the Cedar's primacy.

Finally, it was time to go. Gagarin and Titov, accompanied by a few other cosmonauts and the engineers in charge of monitoring vital functions, boarded a bus specially equipped with seats similar to that inside the capsule to feed the suit's ventilation system. It is said that during the ride to the launch pad Gagarin asked the bus driver to stop because he had to attend to an urgent bodily need. The spacesuit was not designed to do this function comfortably in flight, and he certainly would not want to find himself with his own urine floating around in weightlessness. He relieved himself against a rear wheel of the vehicle.

A habit contracted in the barracks, it then became a propitiatory rite of good luck for all astronauts departing from Baikonur, along with other apotropaic gestures such as drinking a cup of champagne (presumably Armenian), cutting their hair two days before the flight, not watching the placement of the rocket on the launch pad (it also happens with wedding dresses), signing the dormitory door after the last night before the launch, and planting a tree along the Memory Avenue at Baikonur. For his part, Gagarin, respecting a superstitious tradition of Soviet pilots before a major commitment, had not shaved.

As the bus approached the launch pad, Yuri began to see *the silver body of the rocket pointing to the sky, with its six engines capable of unleashing a force of 20 million horsepower*. At the foot of the imposing truss supporting the rocket, where a small crowd awaited him with sober trepidation, his excitement reached its peak. The admiring glances of the onlookers worked much better than the flasks of strong liquor distributed to soldiers in the trenches of the Great War to give them courage before a bayonet attack.

"*This Sun gives the joy of living*", he shouted with melodramatic fervor, or at least that's what's written in his memoirs, which have obviously been skillfully and heavily retouched by the Party's propaganda offices. A statement full of meanings and messages, but far less tragically true than the war cry attributed to the Oglala Lakota war chief Crazy Horse: "*Today is a good day to die*".

After the customary greetings and hugs, Gagarin, codenamed *Kedr* (Cedar), climbed the metal steps to the service ramp elevator that would take him 40 m up to enter the *Vostok*. Awaiting him was a spherical capsule with a volume of 1.6 m^3, considerably less than a Smart car, and a weight of 2.5 tons, mated to a rocket engine to slow it down for re-entry, returning it to the overpowering will of Earth's gravity. It looked a bit like those glass spheres filled with water that you tilt twice to enjoy a snowfall on some famous monument.

A total of 4.7 tons of iron, instruments and fuel to protect the cosmonaut, take him for a walk in the sky, and save his skin. Inside the *sharik* (Russian for "little ball") there were a few commands, which were forbidden to the cosmonaut anyway, and a number of simple instruments for measurement and communication. The climate was typical of Earth, with a pressure of about one atmosphere, an air composition similar to that on the ground, a humidity of 55%, and an average temperature of 20 °C. A real treat, at least in theory.

"*I looked at the spaceship I was about to board for an unprecedented journey. It was beautiful. More beautiful than a locomotive, a steamer, a plane, a palace, and a bridge—more beautiful than all of these creations put together*". A few months before the launch, Korolev had taken a small group of cosmonaut candidates to the factory in Kaliningrad,[2] the "forbidden city" near Moscow, where the *Vostok* was under construction, and had allowed them to inspect the interior of the spacecraft. When Gagarin boarded the capsule, he took off his shoes.[3] For him, "*it was like entering home*", the chief designer would fondly remember.

Seated in the special chair, wired and sealed in his orange spacesuit, the cosmonaut was put on standby. There were several things to do and control before the launch. First the capsule had to be closed. This operation took two attempts due to an alarm on the payload reopening system. Now Yuri was alone, under artificial light, with a camera pointed at his face constantly spying on him, monitored in his vital functions and left to watch instruments he could not even touch.

Unable to predict his reactions during the flight, Korolev had decided that Gagarin would be just a passenger, a luxury guinea pig, without the authority to maneuver the capsule. His main function would be to demonstrate that man can withstand the terrible stress of departure and landing and the hostile environment of extraterrestrial space. The commands would come from Earth, just like any other spacecraft. Only in case of an emergency could the cosmonaut be authorized by the ground station to intervene in the control of the spacecraft. The procedure was protected by a code contained in an envelope given to Gagarin at departure, with which the cosmonaut would unlock the onboard computer.[4]

---

[2] Satellite city of the capital, already a hub of silk processing and, after 1945, the center of the Soviet space industry and related scientific and technological activities. In 1996, it was renamed Korolev in honor of the chief designer, thanks to a memorial campaign initiated by Yeltsin.

[3] Russians are accustomed to taking off their shoes when they enter the house. They ask their guests to do the same, so they keep a decent selection of slippers of all sizes in the foyer. A hygienic rule in a country where snow and mud abound.

[4] The doubt remains whether the authorization to take over the commands in case of emergency was not just a lie told to Gagarin to keep him calm.

"*All is well*", he reported to the control room in a calm voice. His heart was beating normally. In return, the technical director of the flight informed him that the launch would take place in an hour and a half. An eternity! Seraphic, Yuri asked him to send some Russian love songs into his headphones. Perhaps, during the inertia of the long wait, he reviewed his whole life, going back to some spots of an existence that had begun on March 9, 1934, in a modest worker's house in the village of Klushino, Smolensk *Oblast*[5]: a flat region on the upper course of the Dnepr River, today on the border with Belarus, which had seen the invasions of Genghis Khan and Napoleon, and more recently that of Hitler. All of them ended badly for the invaders.

"*I come from a common family*", he would later write proudly, "*a family of workers like there are millions in my socialist homeland. My parents are two simple Russians to whom the October Revolution gave a full and dignified life*". His father Alexei was a carpenter and bricklayer who, despite a severe disability in one leg, mastered "*every kind of trade because he knew how to build with his hands anything that could be needed in the country*side". The son of a very poor peasant, himself semi-illiterate—he had attended only the first two grades of the parish school—and yet curious and ingenious, Alexei was "*strict but fair*". He was a man whose word had the force of law in the family, and whose opinion was highly respected in the Soviet of the collective farm, where he worked with his wife. Anna Timofeyevna Gagarina performed the duties of a milker, taking care of the cows from dawn to dusk. This ensured that there was never a shortage of good milk for her four children, Valentin, Zoya, Yuri, and Boris, and their friends.

Life on the *kolkhoz* (collective farm) was peaceful. The Stalinist purges and their gray terrors did not even touch this agricultural outpost, far from the intrigues and hardships of the big cities. Yuri as Tom Sawyer. A barefoot boy enjoying the countryside and the freedom of the wide open spaces, spying on his older brothers, wanting to be like them and beginning to ask himself the first existential questions. Especially one: what lies beyond the hedge? "*As a boy, I too believed that there was nothing beyond Vesuvius, since I could see nothing beyond it*", Giordano Bruno had written. Instead, Gagarin climbed to the roof of the barn to expand his horizons and dream of flying far away. And when he could, he would ask Uncle Pavel, his father's brother and a veterinarian's assistant, who knew a lot and enchanted his nephews with his stories. He spent whole nights lying on the hay and looking at the sky.

Then, the magic suddenly disappeared. One day in May 1941—Yuri was 7 years old—his father came home from a meeting of the village Soviet with

---

[5] In this same region, in 1919, the famous science fiction writer Isaac Asimov was also born.

a dark face. He spoke to his wife, who *"collapsed on a bench, hid her face in her apron and began to cry silently"*. The war had broken out. For a while, nothing happened in Klushino, except for Yuri's first encounter with an airplane, a Soviet fighter that had landed near the village to pick up a comrade who had been shot down by a Messerschmitt. In November, the Wehrmacht troops marching on Moscow took over the village, which remained under their control for almost two years.

The Gagarins had their home confiscated and were forced to survive in a hastily constructed mud hut. The two older brothers were deported to Poland as forced laborers. They did not return until the end of the war. The father volunteered for the Red Army despite his handicap, and Yuri's younger brother risked being hanged by a Bavarian soldier who *"hated children"*. In fact, Yuri and Boris fought the Germans in their own way, scattering shards of glass on the road, polluting the acids in the batteries, and stuffing potatoes into the exhaust pipes. Twenty-one months of fear, hope, humiliation, and hunger. Then, the Germans began to retreat. They were no longer the arrogant blond warriors of the early days, but tired and demoralized men dreaming of going home. Even Yuri was no longer the same carefree boy. He had to face life and choose a path.

The music in the headphones paused and so did the flow of memories. A voice from the control room asked: *"How's it going, Kedr?"*. *"I hear you well"*, the cosmonaut replied. *"I feel good, mood is excellent, and I'm ready to go"*. *"In one hour"*, said the voice. The communication stopped, the music resumed, and presumably the journey into memory began again in the background. Mind and body were ready to spring into action at the first sign of activity.

In 1946, the family, now reunited, left the *kolkhoz* and moved to Gzhatsk,[6] a nearby town where Yuri completed elementary school. At sixteen, following in his maternal grandfather's footsteps, he became an apprentice in a large foundry near Moscow. He worked during the day and studied at night. To advance, he needed a seventh-grade diploma. Eventually, he was admitted to a special course on tractors at the Industrial *Technicum* of Saratov, a large center on the lower Volga, 800 km southeast of the capital. Tractors were very fashionable. To rebuild the war-ravaged country, Stalin had launched two five-year plans, one in 1946 and a second in 1951 that called for massive mechanization of agriculture.

I continued my studies at the "Technicum", but whenever I heard the roar of an engine in the sky or met a pilot on the street, I could not resist being moved. It was my subconscious passion for space.

---

[6] Renamed Gagarin in 1968, after Yuri's death.

Having learned of the existence of an aviation club and scraping together some money working as a docker on the Volga River, he enrolled and quickly learned to skydive. It was the pre-requisite to start flying. His pilot training took place first on an old biplane and then on a two-seat military primary trainer aircraft produced just after the Second World War, the Yakovlev Yak-18. Yuri had found his way.

"*Kedr, we're starting in half an hour*". Once again, the flight director's announcement had interrupted a nostalgic memory lulled by the languid sweetness of Russian songs. Korolev, whose code name would be No. 20 from now on, intervened to remind Yuri that there was plenty of food in the pantry and to recommend, with the tender humor of a worried father, not to eat too much to avoid gaining weight. Gagarin responded to both and returned to his memories. He could do nothing but wait.

Picking up the thread, he remembered the diploma from the *Technicum*, just six years earlier, the farewell to his flight instructor and the plane that had accompanied him on his first steps in the sky, and the departure for the military service.

> *Many students of the Aeroclub were hired by the civil aviation, attracted by the opportunity to fly across the country or abroad on Aeroflot's now numerous routes. Others had chosen special aviation for agriculture, medicine, geology. As for me, I wanted only one thing: to become a military pilot [...] I liked the discipline and the uniform attracted me. At last I wanted to be a defender of my country [...] I got a letter of introduction to the Orenburg Aviation School.*

Orenburg was taking him further and further away from Smolensk, on the border between Europe and Asia. The military base, located on the outskirts of the city, was named after the pioneer of Soviet aviation, Valery Chkalov, the man who, in the manner of Julius Caesar, said: "*If you must be there, be there first*". Admitted to the pilot course, Yuri immediately stood out for his physical and mental qualities. Of small stature, only 157 cm, but with an irresistible smile, he excelled in everything: in sports and theoretical disciplines, and especially in flying. In Orenburg he met Valya, nickname of Valentina Ivanovna Goryacheva, a medical student with sweet eyes, daughter of a cook, with whom he decided to share his life. At the end of the course, in 1957, they got married. During the party that followed the simple ceremony, prepared by his father-in-law, who "*had sworn to shine in his culinary art*", Valya's mother turned on the radio.

> *We heard these words: "Two ambassadors of the Soviet Union, two stars of peace are flying over the Earth. On the fortieth anniversary of the October celebration, our*

scientists, our builders, engineers, technicians and workers have given us, the Soviet citizens, a wonderful gift by realizing one of the boldest dreams of mankind". We immediately recognized a voice dear to us, that of Nikita Sergeyevich Khrushchev.

A unique coincidence. Lieutenant Gagarin was now qualified to fly the MiG-15UTI, the legendary swept-wing jet that had sown death in the skies of Korea during the recent war. A transonic thoroughbred with a bad temper. A kind of rocket with wings that required courage and determination to tame.

After completing his course, he had to choose his first service destination from a number of options. He opted for a base in the Murmansk region, on the border with Norway. A region with hellish weather. Polar cold, fierce winds, lots of snow and, above all, fog that blinds pilots at low altitudes, where the danger is greatest. *"And why exactly the north?"*, Valya had asked him. *"Because the north is more difficult"*. Five years later, speaking to students at Rice University in Houston, President Kennedy would use the same words to justify the decision to fly to the Moon. A bold and positive approach to life, a *"memento audere semper"* (always remember to dare) that adds the joy of adventure and the taste of challenge to Odysseus' quest for pure knowledge.

Gagarin followed his destiny. By the end of 1959, after an intense flight experience in the icy North, he felt ready for space. Khrushchev had just returned from America and rejoiced at the new success in the space race. Skillful propaganda amplified in the country the merits of the first circumnavigation of the Moon, igniting the spirits of the most daring and ambitious young people. So Yuri made a decision. *"Respecting the military hierarchies, I reported to my commanders and applied for admission to the group of cosmonauts"*. Meanwhile, his first daughter, Elena,[7] was born.

The selection went very well. Gagarin impressed the commission with his physical and intellectual qualities, his incredible memory, great concentration, quick reactions, perseverance, and a strong discipline that did not prevent him from defending his own opinions. He was chosen in 1960 along with 19 other brave pilots. *"Space needed men with a burning heart, a lively spirit, strong nerves, an iron will, high morale, and an unbreakable spirit"*. His wife did not take it well. In addition to the logistic problems caused by Yuri's decision, Valya was worried about her husband's safety. The venture he had embarked on had an equal chance of success and failure: 50 to 50. *"But my father had full confidence in the chief designer, and Korolev loved him very*

---

[7] Today, Elena Yurievna Gagarina is the director of the Moscow Kremlin Museum, where, almost paradoxically for the heir to one of the symbols of the USSR, she looks after the imperial cloaks, scepters, and Fabergé eggs. Icons of a past denied by the October Revolution.

much", his second daughter Galina,[8] born just 36 days before her father's flight into space, would later recall. As the possible launch date approached, Valya's anxiety grew. She had just given birth when Yuri took the plane to Tyuratam. He chose to lie to her, telling her that the mission had been postponed indefinitely.

The *Kedr* was still pondering these thoughts when Korolev informed him that everything was ready for launch. The engines of the first stage of the rocket were fired to preheat, and the giant began to vibrate violently. The noise became deafening, though muffled by the protective headset. "*Pre-stage… intermediate… main… launch! We wish you a good flight. Everything is fine*", almost shouted No. 20. "*Poyekhali*", let's go, was Gagarin's dry answer. At 9:07, the imposing white obelisk lifted off from the platform and began to climb the sky toward the east, faster and faster.

Inside the *Vostok*, Yuri began to feel the weight of acceleration pressing on his chest and locking his limb muscles. He couldn't even speak, but he wasn't worried. He had experienced this feeling many times during training in the centrifuge. After two minutes, at the end of the first stage, there was a moment of respite. "*I'm fine*", he reported. The thousands of people on the huge base, technicians, military, workers, and some distinguished guests, who watched with their noses up and their hearts in their throats, saw almost magically a flaming cross form in the sky, later called Korolev's cross. It was the multiple somersault of the four lateral boosters, now exhausted and detached from the vector.

When the engine of the second stage fired up, the acceleration felt furious anew; and once again Gagarin could hardly speak. Nevertheless, he transmitted all his calmness to the base. "*I feel well. I continue the flight. The overloads increase. Everything is fine*". About 150 s after lift-off, the rocket had passed through the dense layers of the atmosphere. The air was no longer an obstacle. A signal was sent from Baikonur, and some small charges blew up the joints between the two segments that made up the nose of the rocket, under which the *Vostok* was nestled. With the cover of the *Matryoshka* doll removed, sunlight flooded the capsule through the three large portholes and the cosmonaut had a first taste of the view from 100 km up. No one had ever seen it before. "*How beautiful! I was about to shout*", he later recalled. "*But I stopped*". An intelligent and disciplined guinea pig, Gagarin understood perfectly that his mission "*was not to admire the landscape, but to transmit useful information*".

---

[8] Galina Yurievna Gagarina then became a professor of economics at Plekhanov University in Moscow.

The second stage continued to push furiously, positioning what remained of the rocket on a trajectory increasingly parallel to the horizon. Then, the powerful engine shut down, and the boulder Yuri felt on his chest disappeared almost instantly. Four and a half minutes had passed since the launch. For a moment the cosmonaut experienced no gravity. But to reach the goal, one more push was needed to increase the spacecraft's speed to 7.9 km/s and exactly balance the Earth's gravitational pull. That was the job of the third stage. When it too jettisoned at 9:18 a.m., gently separating from *Vostok*, No. 20 breathed its first deep sigh of relief. The *Kedr* was safely in orbit! "*The sensation of weightlessness feels nice. Everything is swimming*".

All had worked, but not perfectly. The spacecraft had settled into a higher orbit than expected, with a perigee of 190 km and an apogee of 330 km above the ground. The timing and logistics of re-entry would have to be re-planned, hoping that the one engine *Vostok* had for deceleration would do the job. In fact, for weight reasons, no provision had been made for a spare. The only precaution had been to carry a seven-day supply of food and air on board, speculating on a natural deceleration of the spacecraft due to friction with the atmosphere. But the trajectory error had wiped out that option. All that remained was to hope for the best. Korolev decided not to worry Gagarin. *Kedr*, for his part, having understood everything, refrained from telling the *glavny konstruktor* so as not to upset him unnecessarily. Later, everyone forgot to inform the world about it, so as not to spoil the story of the heroic deed.

Meanwhile, Yuri was enjoying the view and rattling off exclamations of amazement along with now-famous phrases that reeked of collaboration with the Party's propaganda specialists. Probably prepared statements like this: "*From up here, the Earth is beautiful, without borders or limits. I keep flying. Everything is normal, everything works perfectly, I feel very good. It's beautiful: the Earth is blue*". And others about the presence of God ("*I looked and looked, but I didn't see God*"), which were uttered by others, in particular by Gherman Titov, but which, according to Kremlin experts, sounded good in the mouth of an icon of atheistic socialism.

Over the North Pacific, northwest of Hawaii, *Vostok* entered the darkness of night and returned to daylight after passing Tierra del Fuego. At 10:25 a.m., as it approached West Africa, near Angola, Baikonur sent the command to begin the re-entry procedure. The spacecraft rotated on itself to bring the retro-rocket forward; then, the liquid propellant engine ignited normally. Korolev took another breath. The deceleration lasted 42 s, after which another command exploded a charge to release the now useless engine from the "little ball" Gagarin was in.

The devil, who had been dozing until that moment, wanted to join the game with the tip of his tail. A bundle of electric cables prevented the complete separation of the two units, which began to spin wildly. Gagarin did not lose courage. A few minutes passed and, now over Egypt, the cables finally broke due to overheating from friction with the atmosphere. But the capsule continued its dance, perhaps due to the high symmetry of the structure. With astonishing coolness, *Kedr* decided not to report this to the ground. He was convinced that the unpleasant carousel would have no effect on the successful outcome of the flight and did not want to alarm the base. All this while the acceleration reached almost ten times that felt on the surface of the Earth. At 10:55, the door of the remaining *Vostok* was blown off and the cosmonaut was ejected together with the seat. Still strong accelerations and dizzying falls.

At an altitude of 2.5 km, after releasing the ballast and activating the main parachute opening, Gagarin began his final descent to Earth, as he had done so many times as a civilian in Saratov. In the meantime, he had climbed the ladder. When his adventure had begun, an hour and 40 min earlier, he had been only a lieutenant, but during the flight the Minister of Defense, Marshal Rodion Malinovsky, had promoted him to the rank of major at Khrushchev's request. Beaming as he descended, he whistled to himself a patriotic song set to a tune by Dmitri Shostakovich. The words were by Yevgeny Dolmatovsky, a successful poet of socialist realism: "*The motherland hears, the motherland knows/Where her son flies in the sky*".

A few minutes of slow rocking passed and there was one last incident, the unintentional opening of the reserve parachute, which threatened to inter-fere with the main parachute. Finally, the first cosmonaut in history landed safe and sound in the countryside around the village of Smelovka, in the Ternovsky *rajon* (district), not far from the city of Saratov, where he had learned to fly.[9] It was 10:52, Baikonur time. Shortly before, despite the support of a parachute, the now unmanned "little ball" had crashed on the hard land of Russia, just as Korolev had predicted.

In less than three minutes, Tass issued a press release that would astonish the world. "*The USSR is the protagonist of an unprecedented human triumph*" thanks to the "*Son of October, Yuri Gagarin*". Korolev had insisted to the Kremlin that the news should be given as soon as possible to avoid that the *Kedr*, landing in enemy territory, would be mistaken for a spy. The next day, the official organ of the CPSU, the prestigious *Pravda*, would head-line in large letters: "*SVIERSHILOS*", that is "DONE!"; and the illustrated magazine *Smena* (Change) would title "*Victory! A man in space!*", "*Hello,*

---

[9] In 1991, the landing site was declared a cultural heritage site by the government of the Russian Federation.

space – *says the Russian*". The press took up and expanded the concepts contained in the message sent to the world by the Central Committee of the Party and the Government of the USSR on the same day of the flight:

The Soviet Union has been the first to launch an inter-continental ballistic missile, to put a satellite in orbit, to send a space probe to the Moon, to create the first artificial satellite of the Sun and to launch a space probe towards Venus [...] The triumphant flight of a Soviet man in a spaceship around the Earth crowns our achievements in space exploration.

But in the meantime, the *Kedr*, who had fallen far from the planned landing site, was lost in the countryside. What happened next is part of the legend, adapted for propaganda purposes but also incredibly moving: almost a page from Homer's *Iliad*.

As soon as he touched the ground, Yuri lifted his visor to breathe the fresh spring air that for him tasted of childhood and staggered toward an elderly peasant woman who had come to the fields to weed potatoes with her grand-daughter Margarita. "*I am a friend, comrades, I am one of you*", the cosmonaut shouted, waving at the woman. Unsure of what to do, Anna Takhtarova prudently hid the girl behind her.

Margarita would recall in an interview on *Pravda* a few years later:

I saw this orange monster with a huge head coming towards us. Grandma helped Gagarin take off his helmet – she pressed some kind of button and when we saw a smiling face in front of us, we realized that it was a human being.

Then, the woman offered the cosmonaut a sip of fresh milk in a timeless gesture of hospitality that recalls Homeric customs.

Gagarin, his face beaming, was quickly reached by the military of an anti-aircraft division stationed in the area and immediately taken to the barracks, while it was arranged that the sites where he and the *Vostok* had landed would be kept under surveillance. Freed from his suit, he gave the first informa-tion about himself and the flight to the division commander, posed for a few photos with the garrison soldiers, and then was picked up by one of the heli-copters sent to find him, with paratrooper medics on board. Transferred to Samara, the Cedar was finally able to call Moscow. "*Please report to the Party and Government and personally to Nikita Sergeyevich Khrushchev*", a Moscow Radio bulletin revealed. "*The landing was normal. I feel well. I have no injuries or bruises*".

Meanwhile, the Kremlin began to prepare the scenery for the triumphal entry of the new hero into the capital. Gagarin flew to Moscow on April 14 with a military jet escort. He landed at Vnukovo airport, about 30 km south-west of Red Square, where he was personally greeted by the prime minister. Breaking with protocol, Khrushchev walked up to him on the red carpet with hat in hand, hugged him, and kissed him. The chief designer had to step aside for security reasons and because at that moment he was not functional for the show.

Young, with blond hair, blue eyes, an exotic cut, and a boy-next-door grin under his oversized helmet, handsome in his pilot's uniform with all the medals pinned to his chest, little Yuri embodied the icon of victorious socialism. Thousands of Russians, waving flowers and pictures, framed his ride in an open car to Red Square for the customary parade in front of Lenin's mausoleum. There he received his country's highest honor, the title of Hero of the Soviet Union, directly from the hands of Comrade Secretary.

State television broadcast this moment of authentic exaltation around the world. "*The historical significance became clear to me only on April 14, when we were invited to Red Square and I saw the ocean of people shouting, smiling, all happy, singing songs*", Gherman Titov, Gagarin's backup, also invited to take part in the big party, would explain to a newspaper many years later. "*I realized then that something extraordinary had happened*".

The heroic deed immediately brought Major Gagarin some practical advantages. A car with a driver and a nice four-room apartment in Moscow. Luxuries unattainable for the average Soviet citizen and especially for the Muscovite. After the Great Patriotic War, Stalin had prioritized the recon-struction of industrial sites, using prisoners of war as slaves. Houses, also badly damaged by the war, had to wait. Muscovites lived in *komunalnaja kvartira*, shared apartments, in conditions of serious overcrowding. On the other hand, the city shone with monumental buildings, such as that of the Lomonosov Moscow State University, inaugurated shortly after the dictator's death to celebrate the greatness of the regime. Pompous constructions, of which the metro stations are another glaring example. Closer to the people than his predecessor, Khrushchev had declared a war on excesses, proposing a naked and gray civic architecture, yet economical, so that each family unit could have a roof and a minimum of privacy. But a hero of the Soviet Union deserved much better housing.

In a very short time, Gagarin's popularity surpassed that of the most famous movie stars. People all over the world were impressed by the figure of the heroic cosmonaut with a sunny smile and a natural savoir faire. In just eight months, he received over a million letters, mostly from admirers who declared themselves hopelessly in love with him and ready for any madness. Even Gina Lollobrigida, the Italian film star, managed to sneak among the guests at the reception organized in honor of the cosmonaut during the Moscow Film Festival and, when Gagarin was within reach, she gave him a loud kiss on the cheek. This upset Yuri's wife Valentina. As a good husband, he reassured her: "*It wasn't her [Lollobrigida], Valyusha, but you who accompanied me in the cold of the Arctic Circle. I will be yours until the day I die*". Perhaps the space hero was sincere in that moment, but not truthful.

After April 12, 1961, a date later celebrated as Cosmonaut Day in the USSR (and now in the Russian Federation), the number of newborns named Yuri skyrocketed. "*I always belonged to a very famous family and it's part of my life: I don't remember the time when the situation with my family was different. I can't say whether it's good or bad, difficult or otherwise, it's just the reality*", Yuri's daughter Elena would recall in an interview that became a podcast. And on another occasion the younger daughter Galina added: "*People always treated us with special attention and curiosity, which had its impact on our behavior in public, as we always had to, and still have to, control practically every step and every word*". Complications of princesses' life.

In fact, the first man to ride on an artificial satellite became the poster boy of Soviet propaganda, a valuable ambassador to be used wisely to strengthen confidence in the USSR model among friendly states and to promote it in those countries of the capitalist archipelago where the presence of sympathizers was significant. Major Gagarin was made a member of the Supreme Soviet and sent to Gamal Abdel Nasser in Egypt, to ally Fidel Castro in the Caribbean as president of the USSR-Cuban Friendship Society, and to Western Europe and Japan. Wherever he went, he was received with all honors. During his visit to London, he was invited to lunch by Queen Elizabeth II. This was a great honor that had not even been accorded to Khrushchev, to whom the Queen had offered a simple tea and pastries. The myth of Gagarin (and the Queen of England as well) is also based on a famous anecdote, an episode that, according to the Soviets, happened during the sumptuous banquet at Buckingham Palace.

The table was that of the great events. With so much silverware of all shapes and sizes surrounding his plate, Yuri was in trouble. He had no idea how to handle them, or in what order. Without losing heart, he grabbed the largest spoon and, plunging it into the salad, called out confidently: "*Come on, let's eat Russian style*". The guests looked at each other between astonishment and indignation. Elizabeth, instead, picked up her own spoon and said aloud: "*Gentlemen, let's eat Gagarin style*". Then, Her Majesty whispered to the cosmonaut privately: "*I don't even know how to deal with all this cutlery; it's the waiters who give me the right one*". But that's not all. At tea time, a saucer with lemon slices arrived. Ignoring their use, Yuri grabbed one such slice with his fingers and ate it. Panic gripped the bystanders again. The Queen, impassive, also took a slice and put it to her mouth, concluding the tasting with a "*Just delicious*". And bidding him farewell, she gave her guest two beautiful dolls for his little girls. "*In all unimportant matters, it is style, not sincerity, that counts. In all important matters, it is style, not sincerity, that counts*", Oscar Wilde used to say.

"*He didn't have much time to spend with the family*", Yuri's daughter Galina would later explain, recalling that among her father's many duties was the presidency of the Federation of Water Sports, in which Gagarin excelled. "*But we always went on vacation together, and every Sunday, when he could, we went to the countryside or to visit someone*". The fact is that he could hardly afford to travel as a private citizen because he was "state property" and a victim of his own popularity. Fate had given him the ability to control fear but not to manage success. Noticing that every door opened in front of him, he spent a lot of time using his position as a heroic icon to help friends find a hospital bed or a seat at the Bolshoi. "*Everyone wanted to drink with Gagarin 'for friendship', 'for love' and for a thousand other reasons, and to drink to the bottom of the glass*", General Kamanin had annotated in his diaries.

Under these conditions, it is easy to lose one's head, and Major Gagarin lost his, due to alcohol, luxury, and women. He even went so far as to injure his head by jumping from a hotel window to avoid being caught in the act by his wife while trying to take advantage of the charms of a nurse, and then even boasted about it during a television interview. The *Kedr* was no longer himself. The only way to get him back on track was to let him fly again. But his life was too precious to the system to allow him to frequent military runways and launch pads.

All this, however, was still unthinkable on April 1961. For the Yankee pride the feat of *Vostok-1* was another slap in the face, which the American press did not try to soften in the least. Big headlines in the major newspapers praised Gagarin and his achievement. Although the Russians had again tried to keep absolute silence before and during *Kedr*'s flight, the White House had been alerted while the cosmonaut was still in orbit. A listening station in Alaska had picked up his conversations with Baikonur. Kennedy, newly seated at his desk in the Oval Office, immediately summoned his deputy, Lyndon Johnson, to consider a countermove. By this time, he was determined to win the space race, whatever the cost.

For now, it was imperative to respond adequately to the Soviets' winning longline service, to use tennis jargon. In the absence of anything better, NASA rushed into a ballistic flight of a capsule with a passenger. Launched by a Redstone rocket, the capsule would rise to an altitude of a couple of hundred kilometers and then be brought back to Earth by gravity.[10] Nothing to do with Gagarin, who spent an hour and a half in orbit mocking Earth's gravitational pull, but still a shot that no one except the Russians had ever pulled off. It was the first step of the Mercury project, born in late 1958 to test both the ability to put a human crew into orbit and return them safely to Earth, and the resilience of astronauts to long stays in space.

Alan Shepard (1923–1998), the Air Force pilot selected for the mission, was no longer a young man.[11] He was already 37 years old, half of which he had spent serving his country in war or in difficult and risky activities, testing jet aircraft and flight procedures. The scion of a good family—his ancestors traced back to the passengers on the Mayflower—he had given up a reckless life to be among the candidates for space travel, giving up smoking and chasing skirts to devote himself to intense physical training.[12] He hoped to be the first man to see the Earth from the sky, and when he heard about Gagarin, he got angry and almost broke a table with a punch. But what's done is done. He had to be satisfied with being second.

He left Cape Canaveral at 9:34 a.m. local time on the morning of May 5, 1961, aboard the Freedom 7 spacecraft, twenty-three days after the flight of *Vostok-1*. Before the flight, he had eaten a breakfast of steak and eggs, like a boxer in training. A menu that would become the rule for U.S. astronauts, proving the universality of apotropaic rituals. The launch had been delayed for an hour by low clouds that prevented photographic coverage of the event. At 2.5 min after the "go", the fuel was cut off and the capsule, freed from its ballast, continued its inertial ascent to an altitude of 187 km.

The Americans also have their own anecdote about the physiological needs of astronauts. Given the brevity of Shepard's flight, NASA designers had not considered the possibility that he might need to urinate. But the launch delay had extended his stay aboard the spacecraft to three hours, and the need became urgent. What to do? Alan asked for permission from the Mission Control to relieve himself inside the suit, which was denied. It was feared that the organic fluid could damage the sensors of his vital functions. Finally, his colleague Gordon Cooper (1997–2004) informed him that the electrical

---

[10] Of course, gravity is always present. During the ascent, it consumes the speed, which gradually decreases until it disappears, then it rebuilds it, but in the opposite direction.

[11] The limits set by the Americans for astronaut candidates were more lenient than those of the Soviets: height less than 1.8 m and age no more than 40 years, but technical preparation at the university level.

[12] It is interesting to note the difference in approach to space volunteering between the Soviets and the Americans. Russians were blinded by patriotism, the Yankees worried that a diversion from normal military activities might harm their careers, so much so that they had to be officially reassured on this point.

contacts had been temporarily interrupted and that he could proceed, which Shepard promptly did with great satisfaction. Once again, the Soviets were more pragmatic and far-sighted.

After a total of 15 and a half minutes, including a moment of weightlessness and a final phase of dizzying descent slowed by a giant parachute, the Freedom 7 spacecraft plunged into the Atlantic Ocean 500 km from Cape Canaveral in a protected marine area, where it was quickly recovered by Navy helicopters. Shepard was fine. Like Gagarin, he was essentially[13] a passenger. *"They say any landing you can walk away from is a good one"*, he would later comment, echoing an aviator's mantra.

The flight, which was televised live, was a complete success. Some 45 million Americans watched it and felt somewhat relieved. Among them was John Kennedy, who rushed to use it as Khrushchev had done, proving that the world is a village. He immediately called Shepard to congratulate him and invite him to the White House to celebrate. What better publicity for the young president, who, after narrowly defeating his Republican opponent, had already had to deal with a serious international crisis: the failed attack on Fidel Castro's stronghold by CIA-trained Cuban refugees in Florida. More than a crime, to quote Talleyrand, the invasion of the Bay of Pigs on April 19 had been a glaring mistake, packaged by the head of the Central Intelligence Agency, Allen Dulles, with Eisenhower's blessing, and motivated by the desire to eliminate the Soviet presence from an island so close to the United States. A concern that was fully justified, but handled with an arrogant superficiality that tainted the project of a New Deal in science and space flaunted by Kennedy during his presidential campaign.

Unable to do better, NASA repeated Shepard's feat on July 21 with Virgil Grissom (1926–1967), a thirty-five-year-old pilot and aeronautical engineer who had distinguished himself for his courage during the Korean War. Gus, as he was known to his friends, took off in the Liberty Bell 7 in the same manner as Shepard and with the same result. NASA had done it twice in a row with the ultra-stratospheric shots. But the Americans were still far behind the hated Soviets.

Within a hundred days, three men had violated the sky above the clouds, flying much higher than eagles, where the atmosphere borders the cosmic void. Strangely, all three will die tragically. Gagarin, at a very young age, in a mysterious plane crash; Shepard, at 78, after a long battle with leukemia[14]; and Grissom, in the Apollo 1 fire. Tired of attacking the space pioneers, bad

---

[13] He had briefly experimented with manual attitude control.

[14] Alan Shepard was doubly unlucky because after his first flight he contracted Ménière's syndrome, a disease that causes vertigo and other severe inner ear disorders, and kept him away from launch

fate will bow to Titov, the "Eagle" of *Vostok-2*, to take revenge on him as well, while the cosmonaut, on the threshold of old age, was enjoying the relaxation of his *banja*, the typical Russian sauna. Almost a curse of Tutankhamun.

---

pads for six years. However, he was able to return to space on the Apollo 14 mission and land on the Moon, where he planned to do something eccentric: hit two golf shots with a six iron.

Flag of the Soviets, Lead Us to Victory

Science is not only a discipline of reason, but one of romance and passion.
Stephen Hawking
Everything in space obeys the laws of physics. If you know these laws and adhere to them, space will treat you kindly.
Wernher von Braun

In April 1961, after the prestigious result of Gagarin's flight had been archived, Korolev appeared tired, enthusiastic, and resentful. He had won overwhelmingly. But the knowledge that he was just a number, systematically put aside at public events, did not make him feel better. After so much suffering, efforts, and sacrifice, he thought he deserved a share in the glory of which he was the main architect. But there was no way to get it. Not even Khrushchev seemed willing to make room for him.

Then, he decided to spend a few days on vacation in Sochi, on the shores of the Black Sea. He wanted to unplug for a while and rejuvenate both his mind, which was consumed by worries, and his body, which was weakened by overeating and lack of sleep. But the chief designer could not stop working and kept asking himself what would be the best strategy to follow Gagarin's feat. The thought became an obsession, so he asked Kamanin, the guardian angel of the cosmonauts, to join him at the resort. He felt the need to discuss the matter with someone who could appreciate the various facets of the problem.

Simply repeating the mission would have been a pointless gamble. There was not much to be gained or learned from a second success, and much to be lost from a failure. In order to keep the lead over the Americans, who

M. Capaccioli, *Red Moon*, Springer Praxis Books, https://doi.org/10.1007/978-3-031-54760-7_8

would certainly not wait too long with their countermeasures, it was essential to extend the flight time by multiplying the number of orbits. But by how much? The assessment was complicated by the fact that, due to the daily rotation of the Earth, the vertical line on which the satellite completes its orbit systematically shifts eastward[1] relative to the ground.

At the altitude of *Vostok*,[2] the drift of the landing point implied the exit from the Soviet space after just three orbits. To ensure re-entry within the borders of the homeland, on a safe ground for the cosmonauts but also for people and property (which implied avoiding the west of the country), it was necessary not to exceed three orbits before returning to Earth, or to remain parked in space for at least 24 h. Seriously concerned about the health of his "boys", Kamanin had no doubts. Missions with dogs had shown a criticality in the animals' vestibular system after only three orbits. Going beyond that limit would have been a risk for a human passenger.

But the chief designer thought differently. Accustomed to risks, he felt that the game was worth the candle, even considering the fact that the disturbance of the quadrupeds was temporary and without consequences. The discussion continued on the beach for several days, and finally Korolev made his decision. *Vostok-2* would be a one-day mission entrusted to the reserve superman, Gherman Titov. Undoubtedly an unlucky man. He would have had the hardest test and the lowest reward. That's the way the world works.

One and a half years younger than Gagarin, Gherman was born in Verkhneye Zhilino, in the Altai region. His great-grandparents had moved to this remote land, where southern Siberia borders Mongolia, attracted by the free land distribution program, and there both grandparents had helped found the first peasant commune, the "May Morning". His father Stepan had taken the name Gherman for his boy from Pushkin's story *The Queen of Spades*. A man of good culture, he lived as an elementary school teacher in Nalobikha, the village where the future cosmonaut grew up with a love of literature, music, and also sports, especially gymnastics. A bourgeois childhood, very different from Gagarin's peasant experience.

The myth, which always surrounds heroes, says that from his youth Gherman demonstrated exceptional tenacity and courage. He rescued his fellows who were lost in a sudden snowstorm and guarded a precious sack of flour at the risk of freezing to death. In 1953, when the time came to

---

[1] By how much? It's easy to estimate. The drift at the equator is $40,000/24 = 1670$ km for each hour of the orbital period. At Baikonur, which is located at a latitude of 46°, this is about 30% less.

[2] Necessary clarification because the period is a function of the (maximum) size of the orbit (Kepler's third law) and the latter depends on the power available to the launcher.

choose his path in life, fascinated by the stories of an uncle who was a pilot, he enrolled in the Kustanai Military Aviation School in Kazakhstan.

In 1957, he graduated from the Stalingrad Flying Academy as a jet fighter pilot, and three years later he applied to become a cosmonaut, again without discussing his choice with his wife, who did not take it well either. At the end of the training he was selected as one of the six most qualified for the flight. Finally, after the great illusion and the even greater disappointment, it was time for the great enterprise. On August 4, Titov was officially appointed *Vostok-2* pilot, with Andriyan Nikolayev (1929–2004) as his backup.

Korolev, for his part, had maneuvered politicians and military bureaucrats to quickly approve the new mission, justifying it with the need to verify the degree of adaptation of the organism to weightlessness: a necessary condition to seriously consider a long trip to the Moon. There was only one way to do the test. A human guinea pig had to live long enough in space to meet all the body's needs, from food to sleep to physiological requirements. It was a gamble, because the unknowns were truly daunting. For example, it was not clear what the real effects of prolonged exposure to the deadly solar wind might be, despite our star's relatively quiet phase. The chief designer also planned to test the cosmonaut's ability to manually perform some simple attitude control maneuvers, gradually transforming him from a passive passenger to a pilot. In fact, medical experts doubted that intellectual abilities and reflexes would withstand high acceleration and prolonged weightlessness. A third goal was to produce a film documenting the Earth as seen from space: a fantastic promotional tool with strong scientific and strategic appeal.

Based on the experience of Gagarin's flight, OKB-1, the Experimental Design Bureau, took measures for the necessary improvement of *Vostok*. The television monitoring system and the short-wave communication system between the spacecraft and the bases scattered throughout the big country were upgraded. Korolev wanted to personally supervise the work in the Moscow workshops. The mission strategy, however, remained the same as for the first launch, including the adventurous re-entry by parachute after the cosmonaut was ejected from the capsule.

The choice of an orbit with a sufficiently low perigee to ensure re-entry in case of a malfunction of the single braking retro-rocket was also confirmed. As with Gagarin's previous flight, friction with the atmosphere should have provided sufficient deceleration to bring the spacecraft back to Earth. To cope with this contingency, the air and food supply on board was calculated for a mission of ten days. Again, the golden rule "*melius abundare quam deficere*" (better to have too much than too little) could not be applied, for the usual weight limits. Officially, the schedule called for three orbits of one and a half hours each, which could be extended to 17 if there were no contraindications, for a total of more than 25 h. Emergency landings were possible only during the first six hours, on pain of being forced to leave the country in a land or water area impassable to the Russians.

The launch took place on August 6 at 9 a.m., Moscow time, with the already tested ritual. Awake before dawn, breakfast in space style, dressing, transfer of the cosmonaut and his backup to the base of the powerful launcher, where teams of sleepless engineers had worked all night, and farewell hugs and kisses, real and ceremonial, for the propaganda to follow. No leaks of news were to be allowed until *Vostok* was in orbit, so as not to give the enemy any advantage in case of failure. As the Latin poet Juvenal teaches us in one of his *Satires*, "*Dat veniam corvis, vexat censura columbas*" (the censor forgives the crows and harasses the doves), especially if the doves allow themselves to be uncovered, we add.

It was scorching hot. On the stairs Titov gave his speech, the same one he would have liked to give a few months earlier, when Gagarin stole the scene from him. "*It is hard to put into words the feelings of happiness and pride that fill me. I have been entrusted with an honorable and responsible task*". Then came the dedications and greetings: to the XXII Congress of the PCUS, which would take place in two months, to his "*great friend*" Yuri Gagarin, present at the launch, and to the Soviet government, embodied by the figure of President Nikita Khrushchev. A very different *aplomb* from that of the so-called prayer attributed to Alan Shepard, uttered as the American astronaut was about to be launched into space: "*Dear Lord, please don't let me fuck up*".[3]

Twenty minutes after the launch, Tass announced to the world the entry into orbit of *Vostok*. The news provoked anxiety and admiration of the American rivals. For a while, Titov, codenamed *Orel* (Eagle), enjoyed the spectacle through the portholes of the capsule. The blue Earth dotted with white clouds and the black sky, the stars, and the blinding glare of the Sun. "*Some say God lives up there [in space]*", he later told a Seattle reporter who asked him what he had seen. "*I was looking around very attentively, but I did not see anyone there. I did not detect either angels or gods […] I don't believe in God. I believe in man—his strength, his possibilities, his reason*". Like a printed book!

After the initial enchantment, he began to practice manual navigation of the ship. The *Vostok* could pirouette on itself thanks to the micro-jets of a high-pressure gas released by tangential nozzles symmetrically coupled to produce a twisting effect, useful both for rotating and immobilizing the ship.

---

[3] More likely he said: "*Don't fuck up, Shepard*".

Gherman also began to take the first films of the Earth with a professional camera. Another first!

The temperature on board remained comfortable, between 10 and 25 °C. After a first round, it was time for the publicity that is always the soul of commerce. "*I feel great*". And then a stream of messages of peace and brotherhood sent to all peoples, without forgetting a special mention to Khrushchev. As Socrates said, it is not difficult to praise the Athenians in Athens.

The first test of the vestibular system followed. This was the one that gave the doctors the most concern. At first everything was fine, but only in appearance. At the scheduled lunch time, *Orel* had no sense of hunger. As a result, he limited himself to drinking fruit juice and playing with a drop of liquid that floated in front of his nose. "*I caught it with the cap of the tube [containing the juice] and drank it*". During the fifth orbit, Titov repeated the vestibular tests. Still no problem, except that a certain head movement caused him annoying dizziness. However, he still did not feel hungry. At dinner time, during the seventh orbit, the cosmonaut forced himself to eat some pâté, but was struck with nausea and vomited into a special bag provided for the occasion.[4] The flight controllers panicked. However, the Eagle quickly recovered and tried to distract himself with work. At 19:30 Moscow time, after urinating—the first human being to perform this operation in space and in the absence of gravity, apparently with the appropriate equipment—he prepared to sleep. Like any of us going to bed, he turned off the radio, leaving only the communication channel with the ground station active, closed the shutters of the portholes to block the intense light of the Sun, and fell asleep. A sleep as long as a "*child's*", but discontinuous, interrupted by the strange sensation of having floating arms.

At the start of the seventeenth orbit, No. 20 himself contacted his *Orel* by radio for final landing instructions. Titov packed his bags, carefully stowed the camera and the logbook and relied on the automatic re-entry instruments. Everything went more or less as it did for Gagarin. A sudden braking, the separation of the "*sharik*" from its engine with the already experienced difficulties but this time without the annoying rolling, the flames around the vehicle, the violent acceleration, and finally a little rest. And again the ejection from the capsule, the fall toward a thick bank of clouds, the release of the seat, and the opening of the main parachute.

Passing the clouds, Titov finally saw the Earth. A railroad and a river near the village of Krasny Kut, in the Saratov region where Gagarin had also landed. It seemed that the game was over, but the incident with the two

---

[4] When his companion Nikolaev asked him from the ground station how the flight was going, Titov is said to have replied: "*Andrukha, train your vestibular system!*".

parachutes that had endangered Yuri's life happened again. For Gherman it was even more serious. But with all his skill and coolness he managed to tame the two brawlers and prevent them from fighting. By now almost on the ground, he was grabbed by a strong wind that whirled him around several times and dragged him along after touch down. Farm workers, excited but not surprised because by now every Soviet citizen knew what a man in orange falling from the sky was, ran to him, helped him undress, and drove him to the capsule that had landed a few minutes earlier several kilometers away.

Titov wanted to immediately retrieve the valuable filmed documents and his diary. He was then picked up by the rescue team and taken for the first medical tests. He had been in space for 25 h and 18 min, covering a distance of 700 thousand kilometers and setting an impressive number of records, including that of age. He was and would remain the youngest cosmonaut in history, except for Dutch student Oliver Daemen, who flew as a space tourist on the Blue Origin-operated suborbital spaceflight mission on July 20, 2021, at the age of 18.

In America, frustration and concern continued to grow. The feat was dismissed by NASA with a terse comment: "*Technically significant result*". Instead, it was simply a triumph and was celebrated as such in the USSR. Gherman Titov, also promoted to major in the field, was greeted by Khrushchev at Moscow's Vnukovo airport on August 9, hugged and kissed by the Premier, and taken with Gagarin to Red Square for an exhibition to the people. A big party in the Kremlin followed, with fireworks all over the country. No. 20 continued not to exist for the world, while his Eagle accumulated cash prizes, personal facilities, two orders of Lenin, a medal of the Hero of the Soviet Union, and various international honors (mostly in USSR satellite states).

Like Gagarin, Titov would never return to space. He spent the rest of his life enjoying his fame without too many restraints.[5] He traveled abroad, even to the United States, where he met his American colleague John Glenn, wrote his memoirs,[6] and received a degree in space engineering. Then, there were beautiful women, alcohol and fast cars, prestigious assignments, a job in the Ministry of Defense and a seat in the lower house of parliament, the Duma, by the time hammer and sickle had already fallen along with the Berlin Wall: the *Antifaschistischer Schutzwall* that Walter Ulbricht, the dictator of East Germany, had begun to erect with Khrushchev's blessing just three days

---

[5] Titov began to amaze people when, freshly landed and taken to a barracks for examination, he grabbed a can of beer and, breaking all protocol, drank it eagerly in front of the incredulous doctors.
[6] *I am Eagle*, co-written with the American Martin Caidin and published in 1962, and *17 cosmic dawns*, of the following year.

before Titov's historic flight, to prevent the free movement of people and ideas between the two sectors of the city. At the turn of the millennium, in the year 2000, the Eagle would say goodbye to the world, killed by the carbon monoxide in his sauna, or perhaps by a heart attack. He was sixty-five years old. His enterprise had shown that space was within human reach.

Vostok-2 was the last drop in the bitter cup of Kennedy's first months in the White House. But the young president had already taken steps to counter the unstoppable march of the Soviets. Immediately after Gagarin's flight, he had asked his deputy Johnson to analyze the problem and propose a solution. The energetic senator from Texas had interviewed the leading expert in the field, Wernher von Braun, who had meanwhile made a career for himself. In 1960, NASA had appointed him director of the Marshall Space Flight Center, freeing him and his team of Peenemünde specialists from military control.

Von Braun had long since begun designing a new rocket, a real mule for carrying heavy loads both in low Earth orbit and in deep space. He had named it Saturn to signal continuity with his previous creation, the Jupiter[7] rocket. He had spent over ten years of his new American life convincing people of the possibility and beauty of interplanetary travel, writing popular science books and articles, accepting all kinds of interviews, and participating in television shows. An adept self-promoter, he had associated himself with famous figures in science and entertainment to raise his public profile. Everyone now knew and appreciated him, but not the White House, which was always worried about mixing the devil with holy water.

The ex-Nazi, now an American citizen, made his voice heard when he accepted the directorship of the NASA Space Center. His vision was to move from projects to production. In fact, he was given the go-ahead to build and test the Saturn rocket. In his meeting with the Vice President, he told Johnson that he knew how to put an American astronaut on the Moon (but not on Mars, as Kennedy had naively requested). A lot of time had been wasted, he lamented, and it was not clear if there was enough left to catch up with the Soviets. Certainly, no more could be wasted. The project would require strong political will, a river of money, and a little luck to try to win. Johnson relayed these words to Kennedy.

The solution, Johnson continued in his memo to the president, could be sought in a massive commitment of the nation to a peaceful enterprise aimed at conquering the Moon or at realizing a space station orbiting the Earth. The second option, however, had two strong contraindications. It seemed difficult to deny its potential military use, with all the obvious consequences

---

[7] In Greek mythology, Saturn is the father of Jupiter.

for the delicate international balance in a time of Cold War. And since it was the easiest option from a technological point of view, it was also the most immediate, leaving too little time to catch up with the Russians. More time was needed to get back into the race, and the Moon still seemed a sufficiently distant goal. Kennedy weighed in and decided on the Moon.

So it was that, on a warm spring day in the Oval Office, the seed of Project Apollo was born. To make it a reality, a strategy had to be developed and Congress had to be lobbied for the necessary funding. This would prove no easy task, as it was a reversal of the position Kennedy had taken in his campaign against the alleged profligacy of the Republican Eisenhower administration.

In truth, the soldier president had concluded his second term with words of great wisdom that are worth rereading and meditating on.

A vital element in keeping the peace is our military establishment. Our arms must be mighty, ready for instant action, so that no potential aggressor may be tempted to risk his own destruction. Our military organization today bears little relation to that known by any of my predecessors in peacetime, or indeed by the fighting men of World War II or Korea. Until the latest of our world conflicts, the United States had no armaments industry. American makers of plowshares could, with time and as required, make swords as well. But now we can no longer risk emergency improvisation of national defense; we have been compelled to create a permanent armaments industry of vast proportions. [...] This conjunction of an immense military establishment and a large arms industry is new in the American experience. [...] We recognize the imperative need for this development. Yet we must not fail to comprehend its grave implications. [...] The potential for the disastrous rise of misplaced power exists and will persist. [...] Only an alert and knowledgeable citizenry can compel the proper meshing of the huge industrial and military machinery of defense with our peaceful methods and goals, so that security and liberty may prosper together.

Without wasting any more time, on May 25, 1961, Kennedy, still weak from the embarrassment caused him by the Bay of Pigs debacle, appeared before Congress—with a Democratic majority, it should be remarked—to deliver a speech on "*urgent national needs*".

If we are to win the battle that is now going on around the world between freedom and tyranny, the dramatic achievements in space which occurred in recent weeks should have made clear to us all, as did the Sputnik in 1957, the impact of this adventure on the minds of men everywhere, who are attempting to make a determination of which road they should take. [...] I believe we possess all the resources and talents necessary. But the facts of the matter are that we have never made the national decisions or marshaled the national resources required for such leadership. We have never specified long-range goals on an urgent time schedule, or managed our resources and our time so as to insure their fulfillment. Recognizing the head start obtained by the Soviets with their large rocket engines, which gives them many months of lead-time, and recognizing the likelihood that they will exploit this lead for some time to come in still more impressive successes, we nevertheless are required to make new efforts on our own.

A strong premise, without excuses or pretenses, to arrive at the concrete proposal:

I believe that this nation should commit itself to achieving the goal, before this decade is out, of landing a man on the Moon and returning him safely to the Earth. No single space project in this period will be more impressive to mankind, or more important for the long-range exploration of space; and none will be so difficult or expensive to accomplish.

But why mention the Moon so explicitly? It is worth repeating. Because it was the first and most conspicuous of the victories yet to be achieved by the Soviets, a target still free with which to overturn an order of arrival hitherto systematically unfavorable to America. Everything else had been swept off the table by a disturbing series of winning hands from a mysterious rocket tamer behind the Iron Curtain.

The project presented to Congress was even more far-reaching. A skilled manipulator of the human mind, Kennedy put on the table the futuristic idea of a nuclear-powered rocket to travel the Solar System and declared the firm will of the federal government to support American leadership in telecommunications and weather observation. More or less explicit winks to science, business, military, and economic power, useful to make a wartime drain on the federal treasury digestible.

It was at this point that the president really got to the heart of the matter, with a massive request for money and, more importantly, an oath of loyalty to the cause.

I believe we should go to the Moon. But I think every citizen of this country as well as the Members of the Congress should consider the matter carefully in making their judgment, [...] because it is a heavy burden, and there is no sense in agreeing or desiring that the United States take an affirmative position in outer space, unless we are prepared to do the work and bear the burdens to make it successful. [...] New objectives and new money cannot solve these problems. They could in fact, aggravate them further – unless every scientist, every engineer, every serviceman, every technician, contractor, and civil servant gives his personal pledge that this nation will move forward, with the full speed of freedom, in the exciting adventure of space.

All this in broad daylight, documented by the press, and repeated with even greater force the following year. On September 12, 1962, in Houston, Texas, Kennedy spoke to a crowd gathered at Rice University's stadium to celebrate the opening of the new Manned Spacecraft Center, the Mission Control Center near Galveston Bay, now named for Texan Lyndon B. Johnson.

We choose to go to the Moon. We choose to go to the Moon in this decade and do the other things, not because they are easy, but because they are hard, because that goal will serve to organize and measure the best of our energies and skills, because that challenge is one that we are willing to accept, one we are unwilling to postpone, and one which we intend to win, and the others, too. [...] Many years ago the great British explorer George Mallory, who was to die on Mount Everest, was asked why did he want to climb it. He said, "Because it is there". Well, space is there, and we're going to climb it, and the Moon and the planets are there, and new hopes for knowledge and peace are there. And, therefore, as we set sail we ask God's blessing on the most hazardous and dangerous and greatest adventure on which man has ever embarked.

The gauntlet had been thrown down. The usual creeping work of intelligence was not needed to discover it. Korolev had received the message. He could try to run for cover, aware of the gap in resources between himself and the others. On the one side, there was the American system, with its enormous economic power, robust private industrial fabric, and meritocratic organization based on competition (no matter how unfair, for those who believe that the end justifies the means), on the other side, the chaos of a highly bureaucratized country still under construction, which, in order to maintain its prestige and military power, had long lived beyond its means, indefinitely postponing the realization of some of the promises of socialism. The numerous wits and generous patriotism of so many young heroes were not enough to win this war between Titans.

Who knows if, after reading Kennedy's speeches, the chief designer also thought, as Admiral Isoruku Yamamoto did after the treacherous attack on Pearl Harbor: "*I fear all we have done is to awaken a sleeping giant and fill him with a terrible resolve*". If so, as a skilled poker player, he showed no fear of the opponent's regrouping after the many Russian victories. Perhaps he should have. In fact, "*Democratic and Republican leaders [of the Congress] were generally bipartisan on the future of American spaceflight*", astronaut Alan Shepard later recalled. Like the three Musketeers, all for one, one for all.

Meanwhile, the Mercury project was proceeding as planned, with a launch that would finally put an American into orbit: one of the seven legendary astronauts selected by NASA in 1959, who would also make history on subsequent U.S. space agency missions. The feasibility of the venture had been tested by an earlier launch of the Mercury Atlas 5 spacecraft, which lifted off on November 29, 1961, carrying another chimpanzee, Enos. It completed two of three planned orbits before returning to Earth and landing in the Caribbean Sea with its passenger alive and in excellent health.

As with the previous mission of the other anthropomorphic monkey, Ham, there were several technical problems, including a failure of the machine that distributed rewards and punishments to the animal based on its responses to the tests it was subjected to. The apparent torture, developed during strenuous ground training, was used to force the chimpanzee to show its responsiveness and attention in various flight conditions. But the most serious concern was the rapid depletion of gas for the attitude control thrusters, which are essential for the correct positioning of the capsule at the start of the re-entry procedures.

In any case, the animal had passed the test with flying colors. So NASA, under the pressure of public opinion, which harshly criticized its slowness, decided to risk a human flight. A forty-year-old Air Force pilot with a heroic past and an impeccable private life was chosen.[8] John Glenn (1921–2016) was a typical healthy, athletic, enterprising, and deeply religious American boy, capable of killing to defend his country and of risking his life to glorify it, "*humble, funny, and generous*".

He was launched from Cape Canaveral on February 20, 1962, aboard the Friendship 7 spacecraft, propelled by the usual Atlas intercontinental ballistic missile. The flight was two months later than the "*political*" request to put an American into orbit in the same year as Gagarin and Titov. On the other hand, NASA was between a rock and a hard place. On the one hand a huge rush to catch up with the runaway cyclist, on the other hand the fear of getting a flat tire and falling badly during the chase.

It was the first time a Yankee had been in orbit, and the Soviets' previous experience did nothing but increase the anxiety of their pursuers. The scientific, technical, and managerial knowledge gathered at Baikonur remained

---

[8] Unlike Titov, who was an open atheist despite the spectacle of the cosmos, John Glenn would have said after his second flight on the Space Shuttle Discovery in 1998, at the age of 77: "*To look out at this kind of creation and not believe in God is to me impossible. It just strengthens my faith*".

shrouded in secrecy, much like today's improvements in the engines and aero-dynamics of Formula One cars. No one helped anyone. On the contrary, the rule of misdirection prevailed. The *Vostok* had been exhibited at the 1961 air show at Tushino Airport near Moscow, packaged and disguised with the addition of a set of non-existent fins to confuse the curious Americans. Low blows of a poker game more akin to fox hunting than a card game.

The countdown had been laborious, and Friendship's flight was equally difficult and even risky at one point because of a suspected thermal shield failure. "*I went to manual control and continued in that mode during the second and third orbits and during re-entry*", Glenn later recalled. "*The malfunction just forced me to prove very rapidly what had been planned over a longer period of time*".

The entire team of reserve pilots, Scott Carpenter (1925–2013), Donald Slayton (1924–1993), and Walter "Wally" Schirra (1923–2007) were involved in the ground management of the hectic flight phases. Three orbits totaling nearly 5 h ended with a splashdown in the North Atlantic. Glenn was safely recovered and America rejoiced. On his return to Cape Canaveral, the astronaut was greeted by President Kennedy himself, who led him in a parade in his own car. In New York, he was honored with a ticker-tape parade, a rain of paper fragments thrown from the windows of tall buildings. Nothing like this had been seen since Lindberg's ocean flight 35 years earlier. A grateful nation breathed a sigh of relief, feeling it had caught up with the Russians. "*It seemed that [Glenn] had given Americans back their self-respect*", it was said, "*and more than that – it seemed Americans dared again to hope*".

Glenn's flight was followed by three others, the final acts of the Mercury project. The Aurora 7, launched in May 1962, with astronaut Scott Carpenter on board, who orbited the Earth three times; the Sigma 7, launched in October 1962, with Wally Schirra, a jovial man of Italian descent, son of aerialists, who completed six orbits; and the Faith 7, launched in May 1963, with Gordon Cooper, nicknamed Gordo after one of the chimpanzees that had flown in the early Mercury launches, who stayed on the carousel for 22 laps. Then, the program was shut down because it was understood that it would not have served to catch up with the Soviets. A change of strategy was needed and the Gemini program was born.

Korolev, too, understood that the good times of solitary surprise escapes were coming to an end. In Moscow, instead, enthusiasm for the many victories had created a state of permanent euphoria. Khrushchev demanded "*another proof to show the whole world that the Americans are hopelessly behind the USSR*". The chief designer then submitted an ambitious program to the government: to launch two *Vostok* capsules, one after the other, and let them fly in formation for a long time. A spectacular parade that would have burst the American liver and provided useful information for the hoped-for future trip to the Moon. The approval came directly from Dmitry Ustinov, vice-president of the Council of Ministers and gray eminence of the Soviet defense industry, along with the request that the enterprise be carried out soon, by February 1962, to neutralize the thaumaturgic effect of Glenn's flight.

But Korolev was not ready. He still had to solve a lot of technical problems and fight against the jealousy and excessive caution of some people: more or less flimsy obstacles that were not compatible with the challenge issued by the Yankees.

First and foremost, the High Command of the Soviet Air Defense. The generals with their medal-covered chests didn't care much about the fate of an adventurous pilot, but they were eager to bring back under their control an activity that they considered their prerogative and that involved a huge amount of money. So they set up traps to hamper the path of the chief designer. The head of the cosmonaut team was also against the proposal that the next flight should last several days because he considered it still too risky.

Kamanin was completely honest and had the doctors on his side; among other things, they were worried about the radioactive fallout from the American nuclear experiments. What he wanted was to gradually increase the duration of the flights in order to step by step verify the tolerance of the human organism. It sounded a wise proposal, but Korolev could not afford it, because he knew he would never have the means to make all the flights necessary to complete the tests. The tram that was passing now might not pass again and one had to have the courage to catch it on the run. So he went directly to Khrushchev, who, his goodness, decreed the minimum duration of the flight in three days. It must be said that the authoritative opinion did not pacify the stubborn Nikolai, who continued his quixotic struggle with his head down.

There was also a priority issue that needed to be resolved. The launch pad had been booked to put into orbit some *Zenit*[9] series spy satellites, publicized under the name of *Cosmos* to camouflage their use. These too had been developed by Korolev's team on his OKB-1, probably to curry political favor and continue space activity (the military brass were against it, but were temporarily silenced). Although the vector was the same as for *Vostok*, the first launches went badly, even damaging the ramp, with more wasted money and the usual KGB investigations, always looking for saboteurs. Finally, at the end of July 1962, *Cosmos 7* did its job right and brought back a good haul

---

[9] Originally these satellites were also called *Vostok*, but they were renamed *Zenit* when, with Gagarin's flight, *Vostok* became a name known all over the world.

of images. Now the field was clear for a stunning new space venture with no less than two ships.

An important step was the selection of the team. For *Vostok-3*, the choice fell on the thirty-four-year-old military pilot Andriyan Nikolayev, codenamed *Sokol* (Falcon), already second to Titov; for *Vostok-4*, on Pavlo Popovych (1930–2009), codenamed *Berkut* (Golden Eagle). The final agreement between Korolev and Kamanin provided for a maximum stay in space of three days. Apart from the immense propaganda value of an enterprise that overshadowed all the desperate attempts of the Americans to catch up, there were some tasks to be accomplished and some experiments to be carried out. These included observing the behavior of the second ship and the third stage of the rocket floating alongside, as well as the launch of *Vostok-4* as seen from space. Then, there was the usual photo campaign, a basic attempt at direct communication between the two spacecraft, and even a session of free floating in the cabin to observe the reactions of a man in weightlessness. Activities for which the cosmonauts had been trained for a handful of seconds during the dives of wide-body military jets.

A few days before the launch, the two pilots arrived in Tyuratam on separate flights. Then, Korolev joined them from Moscow and the final tests could begin. The chief designer was more nervous and aggressive than usual. In addition to the growing burden of his permanent anonymity, he was worried about the logistics of a double launch from the same pad within a 24-hour time frame. The second vector had to be armed in record time after the first launch. So the last checks had to be reduced to a minimum. Gagarin was invited to the launch only for decorative purposes. He was almost ashamed of this role, but Khrushchev had ordered that he be pampered, just as Kennedy had arranged for Glenn. The family jewels were to be protected so that they could be displayed when needed.

Vostok-3 was launched on August 11 at 11:24 Moscow time. After a moment of panic when one of the trusses of the service tower failed to detach during liftoff, the giant vector confidently pointed skyward to enter a nearly circular orbit at an altitude of 170 to 220 km. "*And one!*" thought Korolev. At the end of the first orbit, Nikolayev received a phone call from the Kremlin. Khrushchev himself, the "*dear leader*" of Gagarin, congratulated him. It was the usual blessing of power, a kind of battlefield promotion that promoted the promoter. At the top of the command chain of the two superpowers, there was a competition to appear more often and better at the various public occasions offered by the space race. Even today, at the Super Bowl or a Champions Cup final, improvised fans leave the palace to sit in the stands and be seen by the people, showing an interest and a competence they have not, because Paris is still *"well worth a mass"*.

Meanwhile, work was underway at the launch pad to arm the second rocket, which lifted off the following day, August 12, at 11:08 a.m., on a trajectory virtually identical to that of *Vostok-3*. *"And two!"*, No. 20 exclaimed to himself. Due to the perfect setting of the motors of the three stages and the timing of the launch, the initial separation of the two spacecraft was only 6.5 km, which then increased during the subsequent orbits. In fact, the mission did not include any approach maneuvers because it was impossible to control the *Vostok*. The two magnificent probes were now left to the exclusive will of the laws of mechanics and gravity. However, the cosmonauts maintained the umbilical cord of radio contact.

Everything was proceeding so well that the possibility of extending the mission was even considered. Then, the temperature inside *Vostok-4* dropped to 10°, and against Korolev's will it was decided to bring down both the Falcon and the Golden Eagle. It was breakfast time in Moscow on August 15 when the two cosmonauts, 7 min apart and 290 km from each other, landed near the city of Karaganda in Kazakhstan after catapulting out of the spacecraft and opening the parachute according to the now standard landing procedure. *Vostok-3* had been in space for 94 h, or 64 orbits, eclipsing the records of Glenn and Carpenter. One day longer than *Vostok-4*.

The Americans responded with the aforementioned Schirra and Cooper flights, which ended with a great technological show: a live television link with the astronaut in orbit. Small satisfactions for a nation, queen of modern technology, with prestigious universities and research laboratories and with a rich and varied private production apparatus, not yet committed to contrasting the pace of the Soviets in the conquest of space. It was from this fabric of high-tech creativity that one of the most powerful weapons of the Yankees' comeback would emerge: the miniaturization of electronics. The invention of the microprocessor at Texas Instruments, then perfected by Italian-American physicist Federico Faggin at Fairchild, is the most striking example of the combination of science, application, and market that ultimately made the difference between the U.S. and the USSR.

But in that August of 1962, the microprocessor was yet to come, and for the moment Khrushchev was gloating. Korolev, for his part, sneered, for he had another left hook to the chin in store, one of those that stuns the opponent and knocks him out: another tandem mission, but with a sensational *coup de théâtre*. There would be a woman on board one of the two *Vostoks*. A slap in the face to the male astronauts of liberal America, the supermen of aviation forced to give way to simple Russian girls, and a strong wink to women everywhere in an era of rising feminism. *"In my country there is equality between the sexes"*, Gagarin would have told a journalist, twisting the knife in the wound.

The mission, conceived by Kamanin in late 1961 as a propaganda vehicle, had been enthusiastically endorsed by Khrushchev. Suspecting that the Americans were working on the same project,[10] it was scheduled for mid-1963 as the first of seven *Vostok* shuttle launches on which OKB-1 production was concentrated. Unfortunately, there were many roosters in the upper echelons of the Soviet space henhouse, all quite aggressive and eager to grab crumbs of power by obstructing rather than proposing, as often happens in totalitarian regimes and not only. The all-powerful Vice President Ustinov, Leonid Smirnov, head of the Military-Industrial Commission, Kamanin, and many other political and military leaders all wanted a say in strategic and often technical decisions. Some, like Kamanin, did so in good faith; most did so out of ignorance or a thirst for power. This frayed chain of command led to a considerable waste of resources, as several similar projects were often carried out in parallel and unnecessarily in order to feed more mouths with the same morsel. Korolev, aided by the vanity of the Kremlin, tried to negotiate his way through, resisting and conceding, making his blood increasingly bitter.

In the end, in March 1963, during a tumultuous meeting of the Presidium of PCSU, and after a violent attack by the Minister of Defense on the crazy spending on space that was diverting resources from national security programs, it was decided to limit the *Vostok* missions for 1963 to two: one for a long flight of a male cosmonaut and another for a short stay of a woman in space. She had to be healthy, with strong nerves and solid training in parachuting, which was necessary for the landing. The fact that she did not come from the ranks of military pilots and therefore, unlike her male counterparts, had no experience in flying jet aircraft was not a big issue, since the *Vostok* operated automatically and the cosmonauts could be treated almost as simple passengers. At that time, in many countries, it was still considered a gamble to trust a woman with a car, ... let alone a spacecraft.

In March 1962, the selection of possible female cosmonauts began in great secrecy. They had to be women under the age of 30, from all over the country, less than 1.70 m tall, weighing no more than 70 kg, and of impeccable ideology. Of the 400 applicants, only a handful passed the initial tests: their names, with one exception, were kept secret until 1987. These brave women

---

[10] During his visit to America, Titov had been invited to a barbecue by Glenn and, between one beer and another, had understood that the Yankees wanted to use a female pilot in a Mercury mission. Back home, he rushed to tell Kamanin, who used this card to convince the establishment of the need for a woman's flight.

then underwent the usual rigorous training season in the gym, in simulators, and even in the air, in Mig-15s. Not being professional pilots, they also had to learn to fly quickly in order to obtain the ranks and uniforms necessary to save the face of the Air Force.

At the end of the course, Valentina Tereshkova (1937–) and Valentina Ponomaryova (1933–2023) were assigned to *Vostok-5* and *Vostok-6*, respectively. But after the general review of the program, in mid-May 1963 Ponomaryova was downgraded[11] to the second reserve of Tereshkova, who in turn was transferred to *Vostok-6*. The place that had become free on *Vostok-5* was taken by Valery Bykovsky (1934–2019), the usual highly experienced military pilot who had been waiting for his chance for a long time.

Valentina Tereshkova was not only a woman, but also a civilian with no flight experience. Born into a proletarian family in a *kolkhoz* (collective farm) in Maslennikovo, a village in the Yaroslavl region of central Russia, she had a childhood very similar to Gagarin's. Orphaned by a father who died in his tank during the Winter War in Finland when she was only two years old, she started school late and had to abandon it at the age of sixteen to earn a living in the textile industry, like her mother. What saved her from the gray anonymity of an existence under the suffocating dust of the looms was a passion for the void, an unbounded faith in communism, and an iron will. In her spare time, she had begun attending a parachute school at the city's aviation club, joined the Union of Leninist Communist Youth (*Vsesojuznyj Leninskij Kommunističeskij Sojuz Molodjoži*), the *Komsomol*,[12] where she rose to prominence for her activism, and continued studying by correspondence, hoping to graduate.

It was her political fundamentalism, more than any other quality, that made her stand out among the aspiring cosmonauts. She was 26 years old when she boarded *Vostok-6*, ten years younger than Cooper, the youngest American astronaut. Her nomination had been approved by Khrushchev, who saw in Valentina the perfect example of the new Soviet woman, loyal to the cause, upright, and also good-looking: "*a Gagarin in a skirt*". Korolev also approved the choice and planned to use the other more robust candidates for

---

[11] Technically superior to Tereshkova, she was considered less reliable in terms of absolute moral and especially ideological integrity. It is said that when asked what she wanted from life, Ponomaryova answered: "*All that I can have*", while Tereshkova replied: "*I want to support the Komsomol and the Communist Party with all my strength*". And she was sincere.

[12] The curriculum of ideological education of Soviet youth included three levels. (1) the *Oktyabryata*, from October, the month of the Bolshevik Revolution, for students aged 7–9; (2) the pioneers of the whole Vladimir Ilyich Lenin Union (in Russian *Vsesoyuznaya Pionerskaya Organizatsiya Imeni V.I. Lenina*) for boys aged 9–14; (3) the youths of the Communist Youth Union (*Komsomol*) from 14 years. A final level, not accessible to everyone, was membership in the Communist Party of the USSR.

a later flight—which never happened—on a spacecraft with a multi-crewed vehicle.

By late spring, everything at Baikonur was ready for launch. Since June 3, *Vostok-5* had been waiting to take off from the pad now named after Gagarin. Its pilot, Valery Bykovsky, codename *Yastreb* (Hawk), was eager. It seemed that the launch would never happen due to countless last-minute technical problems. Then, the Sun, usually a quiet star, had a powerful sneeze, one of those that send a flood of dangerous charged particles into space. Finally, as the storm subsided, the usual modified R-7 rocket placed the spacecraft into orbit without further delays. It was the early afternoon of June 14.

Due to a malfunction in the last stage of the launcher, *Vostok-5* could not reach the planned altitude. The need to fly in a denser atmosphere, coupled with lingering concerns about the passenger's prolonged exposure to solar radiation, suggested that the mission be terminated after 5 days instead of the originally planned 8. This was still a record for the duration of a single cosmonaut orbital mission. But perhaps the most significant achievement was Bykovsky's absolute acclimatization to weightlessness, which he even found *"pleasant"*. A huge step forward for long trips to the Moon compared to the fears created by Titov's mission.

Meanwhile, Valentina, codenamed *Chaika* (Seagull), had also lifted off, around noon and a half Moscow time on June 16. No problems with the launch. A few minutes later she was in Earth orbit at an altitude of 160–210 km, traveling at a speed of 8 km/s and only 3.5 km away from *Vostok-5*. As before, the distance between the two capsules was destined to increase over time due to the tiny differences in orbital parameters.

The inhabitants of the vast territory of the Soviet Union were able to watch the cosmonaut in her capsule live on television and learned that Premier Khrushchev had called Valentina personally from the Kremlin to congratulate her and wish her a good stay in the sky. But they knew nothing of the serious physical and psychological difficulties and dangers the young Seagull had to face. No. 20 had put a seal of secrecy on the matter, because nothing was to spoil the idyll between the USSR and space. It was an attitude that whitewashed the narrative of the facts and at the same time set traps and pitfalls for historians.

Among other things, the systematic manipulation of information has made plausible the claims of those who, like the famous Italian radio amateur brothers Achille and Giovanni Judica Cordiglia, have accused the Kremlin of hiding some of the disasters collected during the space epic and the list of their victims. First animals, then cosmonauts, crushed on the ground, torn apart by explosions, burned by fire or lost in space. The sad story of Grigory

Nelubov (1934–1966), the third cosmonaut after Gagarin and Titov, who was expelled from the team for drunkenness and a conflict with the military security patrol, is often cited as evidence of the maniacal concealment of dirty laundry. These misdemeanors were incompatible with membership in the fearless and even immaculate team of space musketeers, who were supposed to be witnesses of victorious socialism. After the dismissal, Grigory's name disappeared from documents, and even his image was erased from photographs showing him with his comrades on a group vacation in Sochi. Nothing was to be associated with the heroic conquest of space. A *damnatio memoriae* so ruthless that it led him to commit suicide by throwing himself under a moving train.

The truth about *Vostok-6* came to light after 40 years, when Valentina Tereshkova, now a powerful icon of Russian politics, decided to speak out, thanks to Mikhail Gorbachev's policy of *glasnost* (openness). Her flight, which was supposed to last 24 h, was extended to three days because the capsule was in danger of being shot up and lost in space due to an attitude error. Mission Control made the necessary corrections that brought her back to Earth safe and sound at the end of a long ordeal caused by inadequate clothing, bed sores, and nausea from food. Without losing heart, the girl had tried to spend the time of the 48 orbits sleeping in order to make the air and food supplies last as long as possible, while at Baikonur they tried to figure out what strategy to adopt. "*Chaika has landed*", Gagarin announced to his friend Bykovsky, who was still in space. "*Now it's your turn*". Obediently, *Vostok-5* also returned to Earth.

Another success for the team of the *glavny designer* and another celebration for the people of the USSR, who triumphantly carried the two heroes to the Red Square to receive their dose of decorations. In private, however, Valentina was severely attacked by some Air Force officers. They went so far as to accuse her of incompetence and indolence, and of being drunk at the hearing before the commission that was to analyze her flight. It was even disputed that she had accepted food from her improvised rescuers immediately after landing—made in difficult conditions due to strong winds and the risk of landing in the middle of a deep lake, exhausted, slightly injured, and dehydrated—without waiting for the official rescue team. Tereshkova defended herself with such a determination that it ended badly for her own accusers, who unfortunately did not realize in time which mastiff they were dealing with. Valentina was untouchable.

Gossips reported that there was some tenderness between her and the only bachelor in the cosmonaut team, Andriyan Nikolayev. There had been a hint of it during the goodbyes before *Chaika*'s launch, when Andriyan had hugged and kissed her several times. The rumor had crossed the walls of the Kremlin and reached the ear of Khrushchev, who was always on the lookout for good opportunities to promote his image and the achievements of the socialist regime. It is said that he used Kamanin's good offices to urge the two proletarian heroes, symbols of socialist salvation, to marry. Another clever maneuver to restore a public image tarnished by the failure of the arm wrestling with Kennedy over the Cuban missiles[13] with a pink event of sure media appeal. But also a valuable opportunity for medicine to verify the effects of space on the human body, especially on the reproductive system.

So, on a white November day in 1963, Valya and her Andriyan got married in a grand style in the Palace of Marriages in Moscow. A state banquet followed, attended by the Prime Secretary. The two spouses were given a beautiful apartment on one of the capital's *prospekts* (central arteries) and, almost in return for the many favors they had received, the following year they gave the state a baby girl, Alenka, beautiful and healthy, whose growth would be followed by doctors for study purposes. She herself would later become a successful clinician.

After 17 years of living together, they separated. Andriyan was the classic tribal male chauvinist,[14] more inclined to give his time to friends and vodka than to his wife. In 1982, Valentina married a military doctor, not before obtaining the permission of the new leader, Leonid Brezhnev (when you say fundamentalism!), but always denied that the previous union had been motivated by the sole reason of state. Her loyalty to the ideology and to the CPSU had earned her fame and, in 1966, even a seat in the Supreme Soviet of the USSR. A prominent role for this "*Iron Lady*", which she maintained even after the collapse of the Soviet Union, so much so that she was the flag bearer of the Russian Federation at the Sochi Olympics in 2014.

Forty-five days before the "cosmic wedding", on September 20, 1963, in front of the 18th General Assembly of the United Nations, Kennedy had made an extraordinary offer to the Soviets and the world.

These are the basic differences between the Soviet Union and the United States, and they cannot be concealed. So long as they exist, they set limits to agreement, and they forbid the relaxation of our vigilance. Our defense around the world will be maintained for the protection of freedom – and our determination to safeguard that freedom will measure up to any threat or challenge. But I would say to the

---

[13] When he learned of the intention to withdraw the missiles from the island, Fidel Castro insulted Khrushchev with very strong epithets, while the Cubans sang the slogan: "*Nikita maraquita, lo que se da no se quita*", Nikita naughty (but also gay), what is given cannot be taken back.

[14] Nikolaev was native to the Chuvash Republic, in the center of European Russia, along the course of the Volga River, inhabited by a Turkic ethnic group.

leaders of the Soviet Union, and to their people, that if either of our countries is to be fully secure, we need a much better weapon than the H-bomb – a weapon better than ballistic missiles or nuclear submarines – and that better weapon is peaceful cooperation. [...] Finally, in a field where the United States and the Soviet Union have a special capacity – in the field of space – there is room for new cooperation, for further joint efforts in the regulation and exploration of space. I include among these possibilities a joint expedition to the Moon. Space offers no problems of sovereignty; by resolution of this Assembly, the members of the United Nations have foresworn any claim to territorial rights in outer space or on celestial bodies, and declared that international law and the United Nations Charter will apply. Why, therefore, should man's first flight to the Moon be a matter of national competition? Why should the United States and the Soviet Union, in preparing for such expeditions, become involved in immense duplications of research, construction, and expenditure? Surely we should explore whether the scientists and astronauts of our two countries – indeed of all the world – cannot work together in the conquest of space, sending someday in this decade to the Moon not the representatives of a single nation, but the representatives of all of our countries.

At first, Khrushchev considered returning the invitation to the sender. Why shake hands with your opponent when you are still in the lead? But the idea of saving substantial resources for reinvestment in improving the country's real economy appealed to him. After the almost fatal confrontation over the Caribbean crisis in October 1962, the Russian premier trusted Kennedy, with whom he had reached a reasonable compromise: the dismantling of Soviet intermediate-range missile bases in Cuba in exchange for an equal demobilization (guaranteed on the word) of U.S. ballistic missile batteries in Turkey and Italy. Among other things, the two leaders had agreed to maintain a Red Line of direct communication (first by telex, then by red phone) between them to avoid deadly accidents.

Sadly, two months after his speech at the UN, Kennedy was assassinated on a street in Dallas. On the night of November 23, his vice-president Lyndon Johnson took the oath of office on the Bible and became the 36th president of the United States. But Khrushchev did not trust him, so the proposal for a space alliance fell into oblivion. Another of the tragic consequences of the murderous act of Lee Oswald and those, still unknown, who had armed his hand.

The day before, Kennedy had visited the Aerospace Medical Health Center in San Antonio, 400 km south of Dallas, where he gave his last public speech.

We have a long way to go. Many weeks and months and years of long, tedious work lie ahead. There will be setbacks and frustrations and disappointments. There will be, as there always are, pressures in this country to do less in this area as in so many others, and temptations to do something else that is perhaps easier. But this research here must go on. This space effort must go on. The conquest of space must and will go ahead. That much we know. That much we can say with confidence and conviction.

It had to be that way, but he would not have been able to see it.

The Swan Song

Hasta la victoria siempre (Until victory always).
Ernesto Che Guevara
Success is not final; failure is not fatal: it is the courage to continue that counts.
Winston Churchill

On October 13, 1964, it was very cold in Baikonur. In the control center bunker, the clock showed past one, but no one felt hungry. The anxiety grew as the minutes passed. The last signal from *Voskhod-1* and its crew had arrived as the spacecraft passed over the Caucasus, proving that the descent order had been received correctly. But neither pilot Vladimir Komarov, nor doctor Konstantin Feoktistov, nor engineer Boris Yegorov had shown any signs of life. Korolev knew he was playing the last hand of a true Russian roulette. The new landing procedure had not been sufficiently tested. It's do or die!

Anything could happen, and if things went wrong, the Air Force big shots and the colleagues with whom the chief designer was in constant competition would finally have the winning card to get rid of him and OKB-1. The tension was high. Then came the news that a military pilot had spotted a wobbly object in the sky hanging from a pair of parachutes. Shortly after, the confirmation. The capsule had crashed to the ground in a remote area of the steppe, 300 km north of the city of Kostanay in Kazakhstan. Around it, three people in light clothing waved their arms to attract attention. For No. 20, another strike.

Big hugs, ritual drinks, and immediate communication of the success to Moscow. Everyone waited for the usual fatherly phone call from the "*dear leader*", which never came. No congratulatory message. On the contrary, for

M. Capaccioli, *Red Moon*, Springer Praxis Books, https://doi.org/10.1007/978-3-031-54760-7_9

Korolev and Kamanin came the urgent order from the Kremlin to reach the capital, with the additional note that the usual parade on Red Square had been postponed. For the three cosmonauts who had already flown back to Baikonur, the order was to stay at the base and wait for instructions. Muffled voices whispered that something big had happened in the country's control room. Indeed, a crowned head had fallen, that of Comrade Khrushchev, neatly cut off by a palace conspiracy led by Leonid Brezhnev (1906–1982), President of the Supreme Soviet, and a handful of accomplices.

Khrushchev, vacationing at his Georgian dacha in Pitsunda, on the Black Sea, with his loyal friend Anastas Mikoyan, had received a call from his Kremlin office asking him to return to the capital for an emergency Presidium meeting. Urgent agricultural problems, he was told. Although he suspected a trap,[1] he flew to Moscow without taking any special precautions. It was a fatal mistake or perhaps a conscious surrender, for as soon as he landed at Vnukovo Airport, the KGB took him into custody.

Brought to the Kremlin, Brezhnev, Alexander Shelepin, and the head of the KGB, Vladimir Semichastny, all people who had been close to him for a long time,[2] subjected him to a heavy trial. But no physical violence. The conspirators wanted to make sure that the action, then ironically named "*October plenum*", did not look like a coup. Better to "*do nothing that is not absolutely necessary*", especially since Khrushchev showed no great will to resist. The old lion, worn down by the years, no longer had the determination he had shown at Stalingrad or in the execution of Beria, of which he sometimes, when drunk, boasted that he was the material author. He confined himself to commenting with Mikoyan that if Stalin had been in the saddle, "*not even a shadow would have remained*" of the plotters.

Tired and perhaps frightened, on the morning of October 14 he announced his "*voluntary resignation for reasons of age and health*"—*Pravda* later wrote—settling for a modest pension, less than Gagarin and Titov earned after their heroic ventures, and possession of his residences, the Moscow apartment on the Lenin Hills and the dacha. Brezhnev was immediately elected First Secretary and Alexei Kosygin (1904–1980) Prime Minister. In true Soviet style, he immediately began de-Khrushchevization. So thorough was the operation that even the image of the old leader shaking

---

[1] Before leaving Moscow, Khrushchev had warned the party boyars: "*You are planning something against me, friends. Look here, if something is up I will smash you like puppies*". But then he did not follow up on these threats.

[2] The conspiracy was already in place when, on Nikita's birthday in April 15, 1964, Brezhnev wept and hugged him after reading a message of felicitation from all the members of the CPSU Presidium. "*Yet I well remember/The favors of these men: were they not mine?/Did they not sometime cry, 'all hail!' to me?*", Richard II complains in Act IV of Shakespeare's tragedy.

Gagarin's hand against the traditional backdrop of the Kremlin wall was erased from photographs.

As engineer-cosmonaut Feoktistov, the man who last spoke to the "*dear leader*" from space, later wrote in his memoirs, "*the naive Khrushchev forgot that a dictator cannot afford to relax his grip on the police, the army, and his associates even for a minute*". Machiavelli had expressed this concept more succinctly: "*The armed prophets survived, the unarmed were ruined*". Comrade Secretary was caught naked on the beach. When he died in September 1971, he was already an "un-person", barely worthy of a meager note in Pravda and a burial in the Novodevichy cemetery. But this man of peasant appearance, funny, clumsy, and loud, had made the whole world tremble and hope. "*Nikita Sergeyevich, my father, was loved by some, hated by others*", his son, Sergey, said icily over the grave, choking back tears. "*But he left few people indifferent*".

Thus, eleven months after Kennedy's assassination and just over a year after the death of Pope John XXIII, the third major player in the utopian design to bring peace to the planet and finally end a war that had remained de facto open since 1945 disappeared. Kennedy's observation that "*our most basic common link is that we all inhabit this planet. We all breathe the same air. We all cherish our children's future*", no longer mattered. The Soviet Union had fallen back into the hands of hardliners nostalgic for the Stalinist past, while America was cheerfully heading toward the dissolution of its own image of the Dream Country and getting more and more bogged down in the rice fields of Vietnam.

In conclusion, the three cosmonauts of *Voskhod-1* had left Baikonur on October 12 on "*a calm and icy morning, with the thermometer at −8°, light wind, cirrus clouds, visibility over 20 km, almost ideal weather for a launch*", and had returned, after a flight of only one day, to find a different government and a drastically changed political scenario. A truly spectacular *coup de théâtre* for the first of a new, short series of missions conceived by Korolev's OKB-1. Not even a brilliant Hollywood screenwriter could have done it better.

Let's take a few steps back in time. The mission was conceived in the wake of the success of Tereshkova's tandem. High-ranking aviation officials and some government officials had begun to protest more and more loudly against what they now saw as a mere media circus. "*Korolev works for Tass!*", they complained. There was nothing more to be learned from *Vostok*, nor were there any significant strategic and tactical returns for the Red Army. Nevertheless, the spigot of enormous expenditures for the sterile conquest of space remained wide open, causing a contraction of military investment in national security. The criticism was understandable, but unfair, because it hid the much more real and less noble intention of taking over Korolev's toy, a great generator of success. Unlike the "historic" cosmonauts, now untouchable public icons, the chief designer was an all-too-easy target because of his enduring anonymity. If the Kremlin had left him to his enemies, few would have noticed his disappearance, and many could have benefited from it.

Though on the razor's edge, Korolev had received permission in 1963 to build four more *Vostoks*. But what to do with them if the program was to be canceled? In the impasse of the moment, it was the Americans who unwittingly came to the chief designer's rescue. It was widely known that the Yanks had started a project to launch space shuttles with two astronauts instead of the usual lone navigator. For this reason, they had chosen the name Gemini (twins) for their program.

"*We must do it first*", Khrushchev had thundered, now addicted to the successes of his space team. There was no time to invent something really new to win this challenge. So Korolev came up with the idea of modifying the *Vostok* units parked in the workshop and transform them into minibuses for up to three passengers, one more than the Gemini capsule. A modification that had to be carefully hidden so as not to let the Americans sniff out the moment of technical and political difficulty. It was therefore decided to update the acronym from *Vostok* (East) to *Voskhod* (Dawn), keeping more or less the same sound and the same meaning, since dawn appears in the East.

The first problem to be solved was the lack of space aboard the ex-*Vostok*. To accommodate three cosmonauts where previously there was room for one, not only the internal geometry of the cockpit had to be redesigned, but also the passengers[3] had to be freed from their cumbersome suits and accept the risk of crossing the sky in jeans and T-shirts. It was also necessary to improve the last stage of the launcher to compensate for the increase in total weight to be carried into orbit, and to rethink the landing procedure. The crowded cabin made it impracticable the classic ejection seat procedure in the final phase of the re-entry.

There was no alternative but to leave the cosmonauts on board the *Voskhod* and reduce the speed of the impact on the ground: a requirement that had not considered before, since the capsule descended empty. To this end, in addition to doubling the number of main parachutes, a package of solid-fuel rockets was added which would be fired at the last moment to make the landing even softer. Immediately after touchdown, the passengers would have to get out into the open because within a quarter of an hour, the heat accumulated by the spacecraft as it passed through the atmosphere would have

---

[3] The cosmonauts were also put on a strict diet to keep under control their weight and volume.

raised the temperature inside the cockpit to over 60 degrees, slowly roasting them. Korolev continued to question the validity of this solution and wanted to test it with launches by military aircraft from an altitude of 10 km. He also insisted on running an entire mission of the *Voskhod* uncrewed, and it worked. The chief designer preferred these experiments to long laboratory tests, which were unreliable if incomplete, and in any case too expensive and time-consuming.

Then, it came the question of crew selection. Korolev believed that some of the three cosmonauts could be laymen with different expertise than military pilots: doctors and engineers, for example, provided they met the necessary physical requirements. Given the high degree of automation of the *Voskhods*, a single flight expert on board might be enough. In particular, the chief designer, a sturdy and fatherly man, demanded a place for his engineers to test the machines they had created. A request that was also tantamount to staking a claim to a hotly contested seat. The issue was part of the ongoing conflict between the strategic vision of the *glavny konstruktor* and the prudent approach of Kamanin, who was supported by the military aviation for reasons of convenience. The deputy commander, Marshal Sergei Rudenko, agreed with the idea of broadening the range of qualifications, but demanded an absolute monopoly on the selection of cosmonauts, who, in his opinion, should continue to come from the ranks of the armed forces.

In the end, after exhausting discussions, some low blows, and a futile appeal to Khrushchev's mediation, a seemingly winning trio was formed. But three days before the proposed launch date, two of the members were changed. One because he was the brother of an expatriate in France and the son of an enemy of the people who had been executed during the Great Purge, and another—although this is not certain—because he was Jewish. The resurgence of Stalinism was soon to reappear in the Kremlin.

The new trio was commanded by Komarov, codename *Ruby* (Rubin), a highly experienced and competent engineer-pilot. A thirty-seven-year-old Muscovite, he had had a very difficult childhood, partly because of the war, which had taken away his father, who died in combat under unknown circumstances. Highly respected by his colleagues, who called him "*the professor*" for his knowledge and wisdom, he was a brotherly friend of Gagarin. He had not yet flown, although he was among the most qualified candidates from the beginning, because the chief designer had set a strict age limit for his pilots. Now, with a multi-handed mission, this requirement waived.

Of the two lay passengers, Konstantin Feoktistov (1926–2009) worked as an engineer in Mikhail Tikhonravov's OKB and had participated in the realization of both *Sputnik* and *Vostok*. He would have been the first and only designer to fly one of his own creations. But he did not have the *physique du rôle* or even the party card, which he had never wanted because of his repulsion of Stalinism. Boris Yegorov (1937–1994), a twenty-seven-year-old physician, came from a family of illustrious Moscow clinicians with ties to the Kremlin. It was rumored that his place in the *Voskhod* had been pulled out of a hat as a result of a direct report from the Palace. What to think? Maybe even to fly in space you need recommendations!

The two newcomers received a short training. Only four months of physical preparation and a little bit of astronautics. The ship had to leave, and it could not be kept waiting, despite some unsolved technical problems and a certain atmosphere of abandonment. Korolev was troubled by the thought of his wife Nina, who was hospitalized in the Kremlin clinic for a serious operation. He seemed unusually absent-minded, as did some of the staff, absorbed in preparations for Premier Khrushchev's impending visit to Baikonur on September 24.

Finally, on October 11, after repeated delays to leave the ramp for the *Zenits*, which had military priority, and to repair all sorts of defects and failures caused by the novelties and the hurry, No. 20 decided that "*the rocket could be fueled and launched*". The launch took place in the morning at 10:30 Moscow time, in an atmosphere tinged with fear that something could go wrong.

Without their suits and without the escape hatch, the cosmonauts would have had no chance in case of an accident in the first crucial 45 s of the flight. Fortunately, everything went smoothly and in only 9 min the Voskhod reached orbit at an altitude between 180 and 400 km. Feoktistov immediately realized that the short training was not enough to make the annoying effects of weightlessness acceptable. In any case, the three of them immediately set to work on the planned activities. Mostly biomedical research, fluid testing and coordination of multi-disciplinary activities.

After an hour and a quarter, the United Press International agency broke the news, which was soon picked up by Moscow radio: "*The Soviet Union today launched the world's first passenger space ship with three men aboard, authoritative unofficial sources said*". The cosmonauts did everything they could to attract attention. They told about the wonderful spectacle of the aurora borealis and, flying over Japan, where the XVIII Olympic Games were being held, they sent their ecumenical greetings to all the athletes in the race. From his dacha on the Black Sea, Khrushchev phoned them for the usual congratulatory call. "*Anastas Mikoyan is standing next to me and wants to take the phone from me"*, the leader joked. Little did he know that in a few hours, someone in Moscow would take much more from him.

The landing went fairly well. The spacecraft touched the ground with considerable speed, so much so that *"sparks came out of the eyes of the cosmonauts"*; then, it rolled for several meters, digging a groove on the ground. When it stopped, the passengers were upside down but unharmed. The mission lasted 24 h and 17 min, for a total of 16 complete orbits, equivalent to 700,000 km, as much as the radius of the Sun. Once again, the world applauded in admiration.

With a delay of five days, the new rulers of the Kremlin honored the three heroes on Red Square with the usual liturgy. James E. Webb (1906–1992), NASA administrator from 1961 to 1968, with great intellectual honesty described the feat as *"a major achievement in space"* and a clear indication that *"Russia is continuing its massive space program to achieve national power and prestige"*. All true. But it was now apparent that the Soviets lacked the strategic path that NASA, with a well-defined chain of command and a clear goal, had already conceived to prepare for the final rush.

Yet, despite increasing difficulties, the magician Korolev managed to pull another white rabbit out of the hat. It would have been the last of his career, cut short by an untimely death, and also the last spectacular triumph of the USSR in the race to the Moon, proving the irreplaceable role of the No. 20 in the now long series of stage victories of the Soviets. The *Voskhod* had to be modified to allow a cosmonaut to leave the spacecraft for a spacewalk. The aim was to beat the Americans once again, but above all to experiment with a procedure that would soon be considered essential for the mission to conquer the Moon.

To access space, a hatch had to be designed that could maintain air pressure inside the spacecraft. The reason was not to allow the cosmonauts to breathe, which could have been achieved with pressurized spacesuits, but to prevent overheating of the onboard electronics. In particular, the radio equipment, which was essential for the exchange of commands between the control station and the Voskhod and for voice communication. The lack of air would have prevented the heat generated by the equipment from being dissipated, causing irreparable damage.[4]

The solution was a decompression chamber, similar to those used by divers to enter and exit a bathyscaphe. A cylinder large enough to hold a man, with two watertight hatches at either end: one to allow the cosmonaut to enter the chamber while the outer hatch remained closed, and the second to allow

---

[4] It was precisely the pressure losses that damaged the transmission equipment of many on the Soviet probes to Mars and Venus, for example.

him to exit into the open after sealing the first hatch to isolate the shuttle's interior (a maneuver often used, albeit for different purposes, to gain access to jewelry stores or bank counters). The idea materialized in a pneumatic unit to be inflated in orbit and deflated after use and before re-entry into the atmosphere. A cylinder with an opening of 120 cm and a total length of 2.5 m, made up of 40 canvas air chambers interconnected like a beach mattress.

Another serious problem had to be solved. With the *Voskhod* filled like a sardine can, there was no room on board for the decompression chamber package, carefully rolled up and still bigger than a washing machine, and for a cosmonaut in a spacesuit. The only way to accommodate the new equipment was to sacrifice one of the three passengers. It remained to choose the candidate who would face the terrible test of floating alone in the celestial void, without the reassuring presence of the walls of the spaceship. Obviously, a professional with nerves of steel was needed. Not even the "liberalist" Korolev had any objections on this point.

Taking into account defections for various reasons, of the 20 cosmonauts of the first hour, 7 remained waiting for an opportunity to fly. From this group, two couples were selected to be trained for the roles of commander and "space runner", respectively. After an intensive preparation program and exhaustive consultations among the various actors in a play that was beginning to have too many side characters, it was decided in early February 1965 that Pavel Ivanovich Belyaev (1925–1970) would be at the helm of *Voskhod-2*, while Alexei Arkhipovich Leonov (1934–2019) would perform the extravehicular activity (EVA).

The mission, originally named *Progress*, was preceded by the usual unmanned launch. The dummy flight was necessary to check the functioning of the new equipment that had been developed in only nine months. The experiment did not cost much, since Korolev now had more *Voskhod* units than he could ever launch in the future. The correct opening of the decompression chamber, called *Volga*, was documented by a camera. The success of the test convinced the *glavny konstruktor* that he could risk the manned mission.

One by one, the various commissions in charge of supervising the project gave their approval for the launch. So, on March 18, 1965, at 10 a.m. Moscow time, the spacecraft lifted off from the usual Baikonur launch pad and entered an orbit at an altitude of between 170 and 480 km. After not even one full turn around the planet, the extravehicular activity began. With the help of his companion, Leonov put on the spacesuit. This was an evolution of the suit used by the passengers of the *Vostok*, with a white metal backpack for the oxygen supply. It was to last three quarters of an hour at most. Carbon dioxide, heat, and body moisture would be expelled through a vent valve.

Commander Belyaev then activated the release of the *Volga* tunnel, which inflated within seven minutes, creating a rigid structure with two metal flanges at the ends, each with its lock. Leonov entered the tunnel, closed the hatch behind him, depressurized the ambient, opened the access to space, and slipped into the void. It was 10:30 a.m. in Moscow. A 5-m-long elastic tether was the cosmonaut's only umbilical cord with the spacecraft.

> *I gently pulled myself out and kicked off from the vessel [...] Ink black, stars everywhere and the Sun so bright that I could barely stand it [...] I filmed the Earth, perfectly round, the Caucasus, Crimea, the Volga. It was beautiful [...] It wasn't about courage. We just knew it had to be done.*

The eighth son of a Siberian horse breeder, Aleksei dreamed of becoming a painter. Then, he fell in love with flying, became a lieutenant pilot, and from there an aspiring cosmonaut. Floating in the void, he saw with the eyes of an artist what no man had ever seen from that perspective, slightly dazed by the emotion, the deep silence, and a vague disorientation.

> *It was so quiet I could even hear my heart beat. I was surrounded by stars and was floating without much control. I will never forget the moment. I also felt an incredible sense of responsibility [...] Of course, I did not know that I was about to live the most difficult moments of my life – getting back into the capsule.*

After 12 min, it was time to return to the ship. The Voskhod was about to be swallowed by the darkness of the night. "*Get back, Lyosha!*", ordered Pavel. Abruptly brought back to reality, Leonov realized with horror that his suit had inflated enormously. His hands were floating in the now deformed gloves. The designers had not properly taken into account the pressure difference with the outside. To fit into the *Volga* tube, the cosmonaut took a quick initiative that saved his life. He opened the air valve and partially deflated the suit. In this way he was able to slip into the tunnel and close the hatch behind him. The operation was not easy because he had come back head first, contrary to the planned procedure. However, thanks to his outstanding athletic qualities and the long diets he had undergone in preparation for the flight, he managed to turn around in the tunnel and reach the hatch. Certainly, desperation and the spirit of survival also helped him. Drenched in sweat, he finally reached his seat with a pounding heart and a body temperature that had suddenly risen two degrees, in danger of cardiac collapse.

He immediately received congratulations from Leonid Brezhnev. "*We members of the Politburo are sitting here and watching what you are doing. We are proud of you. We wish you success. Take care. We await your safe return to Earth*". It seemed to be a done deal and instead was the beginning of an unprecedented via *crucis* worthy of an Indiana Jones movie and just as incredible.

Once the maneuver was over, the *Volga* tunnel was no longer needed, and Belyaev blew up the charges for its detachment from the *Voskhod*. The mini-explosion triggered a whirling movement of the capsule, which was difficult to control. Then, the two cosmonauts noticed that the oxygen level in the cockpit had exceeded the safety limit. A spark could cause the spacecraft to explode. But that was not enough. After sixteen orbits and five minutes before the start of the automatic re-entry procedure, the discovery of a fault forced Belyaev to switch to manual control. Time was needed to figure out what to do next, so the pilot decided to stay in space for another orbit, while at Baikonur the mission controllers believed the Voskhod had already touched down. "*Where have you landed?*", asked Gagarin. "*Still in orbit*", was the reply. "*We have to descend manually and we only have enough fuel for one correction. We will land where possible*".

Leonov calculated the route for a landing near Perm, in the Urals, and Belyaev adjusted the capsule's attitude and fired the retro-rockets. Ten seconds after the abrupt deceleration, the re-entry capsule was supposed to separate from the service module. But it didn't. As with Gagarin and Titov, enormous rotations were triggered. A *déjà-vu* that could be recovered with patient waiting. In fact, at an altitude of about one hundred kilometers, the ballast decided to come off. Then the first parachutes opened and finally it seemed that the incredible adventure had come to a happy end. The red-hot ball of *Voskhod* passed through a thick layer of black clouds and touched down almost softly on a blanket of two meters of fresh snow. But not near Perm! The two cosmonauts had landed a hundred kilometers to the northeast, in the heart of a dense forest. "*When will they find us?*", asked one. "*Maybe in three months, with sled dogs!*", his companion replied to lighten the mood. They both knew they were in a big trouble.

When they tried to get out of the spaceship, they discovered that the hatch was blocked by snow and a large tree branch. With brute force they managed to open the passage and get out of the *Voskhod* to breathe the icy air of the good Earth, so pleasant after so many dangers. They were in the Siberian taiga, in the middle of nowhere, in a forest of fir and birch trees, populated by bears and wolves in heat, particularly aggressive. Both cosmonauts were accustomed to the cold, and Pasha Belyaev even had a past as a hunter, but to protect themselves from the very low temperatures they had only their spacesuits and a single rifle to face the beasts. It couldn't get any worse!

But they were lucky. As hope began to fade, their distress signal was picked up by a passing cargo plane, which relayed it. At dusk, they heard the sound of an approaching helicopter and launched a flare. "*We're safe*", they thought, but too soon. It was a civilian aircraft, and the pilot had no way to rescue them because of the dense vegetation. Other planes arrived and dropped off all kinds of comforts for the two cosmonauts, including a bottle of cognac, which of course shattered when it hit the ground, and some very useful leather boots. But as night fell, the curious had to leave and the two castaways were left to fend for themselves. To avoid freezing to death or being torn apart by hungry wolves, they returned to the capsule, covering themselves as best they could. They hadn't even been able to retrieve the parachute cloth that had gotten caught in the trees, and the heating system of the *Voskhod* wasn't working. It seemed that fate was truly against them.

The next morning, they heard the roar of a plane circling overhead. It was snowing heavily. Then some rescuers reached the two shipwrecked men on skis: two doctors, a fellow cosmonaut, and a cameraman to film everything. But there was no way to land a helicopter among the trees. Another night was spent in the open air, this time with some comforts like a hastily built shed, a big campfire, good food, and even a hot bath in a tub parachuted from the sky. The next day, at the end of a nine-kilometer hike with skis on their feet, the helicopter finally took Pavel and Leonid to Perm.

When interviewed, Leonov stated: "*Provided with a special suit, a man can survive and work in space*". He might have wanted to add: "*And he can easily die returning to Earth*", but he certainly didn't say it. That wouldn't have been wise for someone who wanted to live peacefully and for a long time. It almost seems as if this highly decorated superman with a penchant for art had made a pact with death, which he managed to dodge in many different circumstances. First, in an earlier parachute jump. Then in a road accident, when his official car landed in a frozen lake and he heroically saved himself, his wife Svetlana, and the driver. And in 1971, on the occasion of the Salyut-1 disaster,[5] when he was stranded because his partner fell ill at the last minute and the inseparable duo had to give way to the reserves.

But perhaps the most spectacular, though less remembered, event was the tragic ride in the motorcade following CPSU General Secretary Leonid Brezhnev on January 21, 1969. The premier had gone to the Vnukovo

---

[5] The space station was reached in June 1971 by a *Soyuz* with three men on board for a 23-day stay. Everything went well, but at the moment of separation for the return to Earth, the rupture of a *Soyuz* valve caused the total loss of oxygen and the death of the cosmonauts by asphyxiation.

airport with Nikolai Podgorny (1903–1983), chairman of the Presidium of the Supreme Soviet, to greet the cosmonauts of the *Soyuz-4* and *-5* missions, about which we will speak later. He was accompanying them in a procession to Red Square for the usual public celebration when a Red Army deserter disguised as a policeman opened fire on the government car. He missed his target spectacularly. Instead of the hated Brezhnev, the vehicle in the bomber's crosshairs carried Leonov, Tereshkova, Nikolayev, and Beregovoy, one of the new heroes. All of them escaped unharmed, while the driver was killed on the spot. Fate, as we know, has little respect for proletarians.

Korolev, instead, had not made a long-term pact with fate. He had miraculously survived the terrible sentence inflicted on him by Stalin, managing to climb back up the slope of a seemingly marked life until becoming the gray eminence of Soviet cosmonautics, the godfather of the most beautiful race of all times, the space race. But on January 14, 1966, he unexpectedly had to throw in the towel. He had just turned 59 two days earlier. He carried few honors to the grave: two titles of Hero of Labor, received in 1956 and 1961 (he would be posthumously awarded the Lenin Prize and the Tsiolkovsky Gold Medal).

The causes of his death are not known for certain. Sergei was hospitalized on January 5 for what was expected to be a simple operation. He had been suffering from intestinal bleeding for some time due to a polyp that needed to be surgically removed. The official version is that when doctors opened him up, they found a large tumor mass. But why did he die so quickly? Because of a problem that arose during surgery, some think. While he was under the knife, the anesthesiologist tried to intubate him to help him breathe, but the maneuver was hindered by the condition of his jaw, which had been shattered at the time of his arrest. The time wasted on this procedure would have caused irreversible anemia. Still others attributed the death to an immune system weakened by stress and overwork. In fact, six years earlier, Korolev had suffered a heart attack, and on that occasion he had been diagnosed with kidney failure, which would have required care and rest, two words Sergei had long been unfamiliar with. He had other hospitalizations for a variety of physical ailments, including a growing deafness, perhaps aggravated by exposure to the blast of his rockets.

To those who advised him to take care of himself and reduce his commitments, he responded with a sad smile that softened a face increasingly marked by fatigue. He couldn't afford it. He knew very well that Khrushchev would have supported him unconditionally as long as he was able to bring him victory after victory in the successive rounds of the competition with the capitalists. Maintaining this pace of work and success was anything but easy in a country hampered by bureaucracy, the hostility of the powerful military caste, the envy of brilliant colleagues, and a technological and industrial context clearly inferior in quality and quantity to the American one. Nevertheless, No. 20 won, one battle after another, to the point where one wonders if it wasn't all his own work, and only his. Although reason rebels against this cult of personality, one remark is unavoidable. After him, the Yankees, who had always been at the back, took the lead in the race.

The head of OKB-1 was taken by his deputy Vasily Pavlovich Mishin[6] (1917–2001), a competent engineer but a great vodka drinker, arrogant and at the same time *"hesitant, uninspiring, poor at making decisions, over-reluctant to take risks and bad at managing the cosmonaut corps"*, according to Leonov. Probably the right character to live in the shadow of a giant and yet inadequate to carry on the project that Korolev, dying, had left unfinished: the conquer of the Moon before the Americans.

Let's take a step back. The idea of a manned space mission had been circulating in the offices of the OKB-1 since 1959, when humans had not yet ventured into orbit. While most of the designers of Korolev's bureau were busy with the realization of the first orbital flights of living beings, a group of dreamers, following the example of Tsiolkovsky, practiced imagining a flight of three men to Mars to perform a close passage around the planet and then return to Earth. A three-year voyage aboard an interplanetary probe launched by a colossal rocket 120 m high and 20 m wide at the base. The celestial ship would carry a large living space, doubly shielded to protect the passengers from cosmic radiation (for what was then understood of the problem), made comfortable by artificial gravity. The preliminary project, signed by Korolev and authoritatively approved by the Estonian physicist and mathematician Mstislav Vsevolodovich Keldysh (1911–1978), President of the Academy of Sciences of the USSR, was submitted to Khrushchev, who did nothing with it.

Korolev did not give up. Despite a full-time commitment to the development of *Vostok*, he returned on April 12, 1960, exactly one year before Gagarin's flight, to drum up support for an even more ambitious project. A landing on both Mars and Venus, to be carried out in a formation of three or four spacecraft, each with a total weight of 40–50 tons. More than the engineering of the carrier and the probes, the document illustrated the launch and return strategy, based on a first stop in low Earth orbit, a hub to assemble the parts of the spacecraft sent with multiple launches, and on the return, the

---

[6] In 1974, after repeated failures in the management of the *N1* mega-rocket project and disappointment at being overtaken by the Americans in the space race, he was replaced at the head of OKB-1 by Valentin Glushko.

recovery of the resources parked there. Proposals were also made for nuclear-powered engines and vehicles for moving astronauts on the surface of the Red Planet. Real trains of platforms with various functions, from mining activities to housing for cosmonauts, to the production of energy necessary for the mini-colony.

Unfortunately, the unmanned exploratory missions sent to Mars in 1962 all failed for various reasons, with serious economic and image damage: three flights of the Mars probes powered by the *Molniya* vector, of which only one managed to leave the Earth and then got lost in the nothingness of the Solar System. A terrible omen for a project that was beginning to appeal to many, in a field where confusion and improvisation now reigned supreme.

Paradoxically, there were too many thinking heads and an exorbitant number of ambitious programs, but they were conducted without adequate coordination and economies of scale. There was a lack of a single, clear strategy, a stubborn determination to serve too many masters, and a disconnection between the demands of science and the needs of the military, which controlled the purse strings. Also notable was the conspicuous absence of the Academy of Sciences, which focused heavily on issues of ideological integrity, while an endemic lack of funding forced builders to take often deadly and ultimately uneconomical shortcuts. For even in space, *"you get what you spend"*.

In 1962, the Kremlin, alarmed by Kennedy's claim that the Americans planned to land on the Moon within the decade, called for a general review of the strategy for attacking space. *"You don't want to leave it to the Americans!"*, thundered Khrushchev, who for the moment would have been satisfied to see a Soviet probe with a human crew fly over the Earth's satellite first.

The following year, Dmitri Ustinov, who had become vice-president of the Council of Ministers, proposed that the missions to Mars and Venus be repeated. But, he said, they should start after a battery of tests beyond Earth's orbit to verify the functioning of the communications systems that had caused the most problems. For reasons of economy and efficiency, the designers were asked not to constantly reinvent the wheel, but to recycle and adapt the components of the vectors and probes already tested and working.

The head of OKB-52, Vladimir Nikolayevich Chelomei (1914–1984), seized the opportunity and in early 1964 proposed a lunar flyby mission based on the *UR-500 Proton*, a rocket his office had nearly completed. A super ICBM designed to carry a 10-megaton atomic bomb over long distances, it used a fuel composed of toxic liquids that would ignite on simple contact. The mission required less thrust, simpler equipment, and most importantly, it eliminated all the problems associated with landing a human crew on the

Moon. In fact, the strategy was to send a single cosmonaut to our satellite on a spacecraft renamed *Lunniy Korabyl* (Lunar Craft) and bring him back to Earth after orbiting the celestial body. Had the project succeeded, it would have created a media impact capable of satisfying the Kremlin's ambitions and overshadowing the glory of Korolev's achievements, thus nibbling away at some of his power.

However, Chelomei did not devote enough time to this project, as he was involved in the development of the *Almaz* (Diamond) program, aimed at putting into orbit space stations for espionage.[7] A strong figure, much listened to in the Kremlin, until that moment he had been mainly engaged in designing vectors of military interest, earning the esteem and gratitude of Khrushchev and the higher echelons of the Ministry of Defense.

He was born in the small Polish town of Sedletse, then part of the Russian Empire, two days after the assassination of the heir to the throne of Austria-Hungary and his wife in Sarajevo. But he had spent very little time there. As the front of the Great War approached Sedletse, his family had decided to move to Poltava, in southeastern Ukraine, where he grew up in a culturally privileged context, in communion with the heirs of Gogol and Pushkin, learning to play piano and appreciating the reading of the classics. At the age of 18, he entered the aeronautical department of the Kiev Polytechnic Institute where he stood out for his vivid intelligence: a model student, if not a precocious genius. In 1940, while Korolev was rotting in a *gulag*, he was selected as one of the top 50 students in the USSR for a special doctoral program that Stalin wanted, with a princely salary. Vladimir was the youngest of the group. Young, yes, but also brave, because he dared to refuse Beria's invitation to become a resident member of Soviet espionage in Germany. After obtaining his doctorate in 1941, he found work at the Central Institute of Aviation Motors in Moscow, where he earned his place by promising a hand in preparing his director's doctoral dissertation. A sign of a transversal *modus operandi* (way of doing things) that would accompany the entire career of this aerospace giant.

Three years later, a German rocket, a V1, captured intact by the British and sent to Stalin by Churchill, arrived in Moscow. In those difficult times, the two bitter rivals, allied against the common enemy, exchanged useful favors. One day, Malenkov, the deputy in charge of the aviation industry, asked "Professor Chelomei"—so he had pompously written on his door—what he thought of the idea of copying the weapon. The answer was so convincing that the young engineer soon found himself at the head of a

---

[7] Three of these stations were actually launched by the USSR between 1973 and 75 under the name *Salyut*, meaning "salute", to disguise a military enterprise with a civilian acronym.

special design bureau called OKB-52, located in a suburb of Moscow. The office immediately distinguished itself for a wealth of ideas and some concrete war applications, including a cruise missile similar to the V1.

In April 1944, Stalin himself summoned him to give his opinion on the feasibility of the project to destroy Berlin with a massive bombardment of V2 rockets. Chelomei had the audacity to contradict the dictator, pointing out that the consequences could have been catastrophic for innocent civilians and for the monuments of a beautiful city. Apparently, he sounded convincing this time as well, because Stalin gave up on the idea and he was not sent to a *gulag* in Siberia. But why did he adopt this moderate position? He probably knew that his rockets, so similar to the V1, would do very little damage to the enemy, and that would have made him look very bad in Stalin's eyes. In fact, he almost ended up in prison when his rocket lost out in the competition with Korolev's project, which had instead replicated the V2. He was fired from his job, which went to Artiom Mikoyan (1905–1970), chief designer of the famous MiG fighters, who had also played the trump card of giving a job to a son of Beria, Stalin's security chief. Chelomei had to settle for a teaching position at the Moscow Higher Technical School.

Shortly thereafter, the dictator died and Malenkov, who had taken over the reins of the country for a time, remembered the brilliant young engineer and put him in charge of a special design group, SKG-10, based at Tushino near Moscow. There, Chelomei developed a winged missile, particularly suitable for use in submarines. In 1955, the SKG-10 was again renovated at the *NPO Mashinostroyeniya* (Research and Development Bureau of Machine Manufacturing) and, eleven years later, OKB-52, a Missile Design Bureau associated with a large industrial complex in Reutov, a town on the eastern outskirts of Moscow. *Alter ego* and bitter rival of Korolev and his OKB-1, Chelomei again had to compete with other giants of Soviet aerospace, Mikoyan and Ilyushin, but this time his underwater missile project prevailed. Then, he changed the subject of his interest and devoted himself to intercontinental ballistic vectors, both for military purposes and as launchers of space probes. He also wanted to anticipate and secure a political parachute by giving a job to Sergei Khrushchev, the son of the then Premier. He wanted to play on an equal footing with Korolev, and he succeeded.

But the chief designer was a tough nut to crack. As early as 1962, more in reaction to Chelomei's intrusiveness than out of any real interest, he had submitted a new and very clear proposal for a multi-purpose launcher called *N1*. It would become the only unfinished symphony of this extraordinary composer of space melodies. The *N1* was to be used for five types of missions to the Moon: a simple flyby (code *L1*), the landing of an unmanned self-propelled vehicle (*L2*), a human landing on the lunar surface (*L3*), the establishment of a futuristic space station that never got beyond the conceptual stage (*L4*), and the sending of a rover piloted by a cosmonaut (*L5*). Mostly hats on the chair. There was neither the time, nor the economic resources, nor the industrial context to do all this.

With the fall of Khrushchev, the struggle resumed, especially since Ustinov's open hostility to Chelomei paved the way for Korolev. "*When talking about the Russian space program, there is a misconception in the West that it was centralized*", Sergei Khrushchev would have clarified without mincing words in an interview for *Scientific American*. "*In reality, it was more decentralized than in the United States, which had one focused Apollo program. In the Soviet Union, there were different designers who competed with one another*".

In the competition between the various bureaus, the leaders of the rocket industry wanted to finally take command, while the Army continued to demand that the entire civilian space program be scaled back. Everyone knew very well that any project, no matter how powerful and innovative, would always be at the mercy of the whims of politicians, the arrogance of generals, the envy of colleagues, and good or bad luck. These kinds of obstacles exist to a greater or lesser extent in every country, but they were particularly effective in the USSR because of the strength of the military, the power of the *nomenklatura*, and the impressive number of brilliant scientific leaders who were all eager to be in the limelight.

In particular, the discussion focused on the vector that would launch the Lunar Module. Korolev's *N1* or Chelomei's *UR-500 Proton*? A special commission of experts chaired by Academician Keldysh was charged with untangling the knot. It consisted of representatives of the Ministry of Defense, design bureaus, and the Academy of Sciences. Most of the members supported Korolev. Keldysh leaned toward Chelomei. In the end, a compromise was reached, and in August 1964 the government issued the following directive. The OKB-52 (Chelomei) would take over the *L1* program for a lunar flyby and the OKB-1 (Korolev) would be responsible for the *L3* mission for landing on the Earth satellite, using a shuttle that his project office had been studying for several years, the *Soyuz*—a word that in Russian means Union.

On October 25, 1965, now under Brezhnev, there was another change of course. Probably the Kremlin preferred to invest in the old route, given the overabundance of new projects. Korolev brought home the whole lunar project and offered a downsizing of the *Soyuz* shuttle as an economic alternative. But he was furious:

> *The Americans have unified their forces into a single thrust, and make no secret of*
> *their plans to dominate outer space. But we keep our plans secret even to ourselves.*
> *No one has agreed on our future space plans – the opinion of OKB-1 differs from*
> *that of the Minister of Defense, which differs from that of the VVS [air force],*
> *which differs from that of the VPK [military industrial commission]. Some want us*
> *to build more Vostoks, others more Voskhods, while within this bureau our priority*
> *is to get on with the Soyuz. Brezhnev's only concern is to launch something soon, to*
> *show that space affairs will go better under his rule than Khrushchev's.*

The immortal Chelomei did not give up and was finally able to rejoin the game with a deal that nobody really liked[8]: the launcher for the low parking orbit would be his Proton, with a fourth stage for travel to the Moon to be borrowed from the N1 super-rocket being studied at OKB-1. On December 31, two weeks before Korolev's death, the two chief designers signed a truce. The third cockerel in the henhouse, Valentin Glushko, out of resentment against his old friend Sergei, did not want to take part in it and preferred to divert his skills to something else. Another symptom of the deadly anarchy.

The pact called for 11 lunar flybys (Korolev's *L1*). The first seven had to be unmanned to test the various subsystems. Such a program required the choice of total spacecraft automation: a strategy constantly advocated by the *glavny konstruktor*, who believed he could not do without it. Indeed, he had his good reasons. But leaving complete mission control to the onboard computers and ground stations presented some serious drawbacks. It complicated the hardware and humiliated the crews.

In any case, this grudging agreement resulted in a powerful and durable mini spacecraft that is still in use today. Seven meters long and almost three meters wide, with a habitable volume of 10 cubic meters and a launch weight of 6.8 tons—which became less than half that at the time of descent—the *Soyuz* consisted of three interconnected elements. At the rear was a service module with retro-rockets, attitude thrusters, fuel reserve, communications antennas and solar panel hooks. In the middle, an acorn-shaped re-entry module designed to protect the cosmonauts during both launch and descent, with a periscope, porthole and parachute compartment. Finally, at the front, an orbital module that served as a living room during the flight. Similar to a big ball, it carried the docking mechanism for connection with another

---

[8] Despite the competition between the two, Korolev respected his opponent. "*Do not underestimate Chelomei. He is of the same design school as Tupolev and Myasishchev. If we give him the will and the means, his products will equal those of the Americans. Now is the right moment to combine forces with Chelomei*".

unit, the door for the space walk, the pantry with tube delicacies, the equipment for monitoring and flight control, and the equipment for scientific experiments.

The first unmanned test took place in November 1966, almost a year after Korolev's death. A *Soyuz* was to be placed in orbit and wait for another shuttle for an automatic docking attempt. But after the first spacecraft was launched, it lost attitude control and began a natural re-entry. When it was clear that the object would fall on Chinese soil, it was blown up in flight to avoid a case of "pitch invasion" and to prevent the delivery of classified technologies to a no longer friendly nation. The failure was masked by the usual technique of naming the mission *Cosmos* (No. 133 for the record) in order to eliminate any reference to the missed targets.

The next test, the following month, went even worse. Like horses before a barrage, the rocket refused to take off, and while technicians were defueling it to defuse the danger, it exploded on the launch pad, causing one casualty and extensive damage. Finally, on the third attempt, on February 7, 1967, everything went well until the re-entry. The spacecraft crashed into the icy surface of the Aral Sea, broke through the ice sheet, and plunged into the water. It was recovered with great difficulty by divers.

There was little to celebrate. Instead of worrying and taking appropriate countermeasures, Mishin agreed to launch a manned flight in early spring. The push to take the risk came directly from the Kremlin. After three years in power with little to show for it, and with so much money invested and systematically thrown back at him by his generals, Brezhnev demanded to celebrate a great achievement, the kind that had made Khrushchev gloat. The time he chose for this was perfect: 1967 marked the October Revolution's 50th birthday, and April marked Lenin's birthday. A new Soviet victory in the space race might possibly bring the peoples of the vast and heterogeneous archipelago of the Warsaw Pact back under the shadow of the Red Flag and make them forget the aberrations of the Prague Spring, which had started on January 5 with the rise to power of Alexander Dubček.

The Americans were mourning the loss of so many of their young men and the embarrassment caused to the world's most powerful nation in the ongoing war in Southeast Asia. On the night of January 30, the Vietnamese New Year, General Giáp had launched a deadly offensive by the northern troops and the Viet Cong against the cities of the south and the American bases, barely repulsed by the Green Berets and with brutal methods. The turmoil of an America judging itself reverberated everywhere, reaching to scratch the marbles of the White House, to the point that Lyndon Johnson decided to withdraw his candidacy in the Democratic primaries for the presidency of the United States, inadvertently launching the Republican Richard Nixon.

Grief after grief, the Americans also mourned the three astronauts who died in the Apollo 1 accident. It was a terrible start for the program to put the Yankees on the Moon, which for a moment nullified the psychological advantage gained with the chain of Gemini successes, which we will examine in the next chapter, and reopened the terms of the competition. In short, Mishin thought, now or never! That was the way his master' acted, but only Korolev knew how to turn a gamble into a victory.

The mission designed by Mishin involved the exchange of passengers between two spacecraft in Earth orbit, with the following strategy. A first *Soyuz-1* would have launched from Baikonur with a single passenger, veteran cosmonaut Vladimir Komarov, who had already flown on the *Voskhod*. The next day, from the same pad, *Soyuz-2* would have blasted off, absolutely identical to the other, but with a crew of three cosmonauts. In addition to commander Valery Bykovsky, two rookies, Yevgeny Khrunov (1933–2000) and Aleksei Yeliseyev (1934–), were chosen to perform a highly spectacular task. After the two spacecraft met and docked, they were to go out into space and transfer to Komarov's *Soyuz-1*. After the exchange, the *Soyuz-2* would have returned with the only pilot on board, and the next day the *Soyuz-1* would also descend with three passengers. This was not just a circus stunt, one of those that provide a cathartic adrenaline rush to the spectators for the meager price of a ticket, but rather an important test of an essential maneuver for a lunar flyby in the absence of an internal communication route between the two spacecraft.

In the weeks leading up to the launch, control engineers at the *Sojuz* had found over two hundred more or less serious problems and bugs. The risks to the success of the mission and the very lives of the cosmonauts seemed to everyone excessive, except to Mishin, who did not feel like displeasing the Kremlin. It is true that "*audentes fortuna iuvat*" (fortune helps the bold), as Korolev well knew, but it is also true that "*est modus in rebus*"[9] (there is a measure to things). In short, one has to take risks but knowing how to handle them.

Gagarin, a close friend of Komarov and his second on this flight, desperately tried to postpone the launch by writing a long memorandum to the premier himself. Fearing that his petition would get lost in the Kremlin maze, he entrusted it to a high-ranking KGB officer, Venyamin Russayev, who was also a good friend of his, to deliver it to the secret rooms of the palace. But,

---

[9] The first Latin quote is from Virgil, the second from Horace.

even if Brezhnev read the document, he showed no sign of understanding the gravity of the situation. According to recent mythology, Komarov even tearfully confessed to Russayev that he knew he would not return from this mission alive, and yet he could not resign because of his duty to protect his backup, Yuri. "*He'll die instead of me. We've got to take care of him*".

On April 23, 1967, the day of the launch of the first of the two *Soyuz*, there was a general atmosphere of despair among the cosmonauts gathered at Baikonur. When Komarov was brought by bus to the usual ramp from which all Korolev's creations had launched, he looked like a condemned man, resigned to his fate. His companions tried to cheer him up by singing classic Russian songs, and the therapy worked. Vladimir smiled as he entered the spacecraft.

At 3:30 a.m. Moscow time, the rocket lifted off the pad and headed for the first shadows of sunset. Eight minutes later, Ruby was in orbit, testing a very sophisticated machine, the most complex ever put into the sky and, for the moment, the least safe. Down at the Baikonur control center, there was not the *glavny konstruktor* to guide the operation with his magic hand. The problems started immediately. One of the solar panels refused to open, cutting electrical power in half. Then came other problems with attitude, thermal control, and communications with the base. At Baikonur, they were considering repairing the faulty panel with the *Soyuz-2* cosmonauts in extravehicular mode—it would have been an extraordinary plus for the mission, a first for workers in weightlessness—when a terrible storm of water and wind hit the base.

It seemed a good reason to cancel the second launch and bring home the cosmonaut in flight, especially since the batteries, lacking power, were about to run out. Considering what was about to happen, it almost seems that *Iupiter Pluvius* (Jupiter bringing rain) had decided to save the skin of Bykovsky and his companions. Death, instead, played with Komarov like a cat with a mouse, with a series of paws, all of which the cosmonaut skillfully dodged, except for the last one. The drama began when the mission control center gave the re-entry command to the automatic system, which refused to work. Remembering the precedents of Belyaev and Leonov, Ruby let another orbit pass before switching to manual control, which he had not had a chance to practice. Nothing tragic so far. If anyone could tame the *Soyuz*, it was him.

The descent procedure required the spacecraft to be rotated to position the retro-rockets in the direction of travel in order to slow the race down and leave the orbit. Despite the attitude control failure, Komarov, who had ridden many angry bulls, was undaunted and rolled up his sleeves. After painstakingly achieving the desired alignment, he fired the engines, which shut down earlier than expected. No panic! It had happened before. That would imply a longer downward trajectory and an off-target landing, almost as scripted. The dive to Earth began badly, however, as the spacecraft, made unbalanced by a crippled solar panel, started to spin on itself as it entered the atmosphere. Without the possibility of maneuvering, Komarov was unable to tame it. The *Soyuz* came down like a top. Nothing irreparable so far. But when the spacecraft reached the denser layers, the main parachute refused to come out of its housing and the reserve parachute did not open either. Without any brakes, the *Soyuz* fell in free fall like a 3-ton meteorite and crashed into the steppe of the Orenburg region in the Urals at 6 a.m. (UTC) on April 24. The rockets designed to cushion the impact with the ground exploded, incinerating the poor remains of the passenger.

Komarov remained conscious and aware to the end. The legend built around the tragedy, which has been the subject of much speculation, offers two opposing scenarios: that of a hero who faced death with determination, spoke to Prime Minister Kosygin on the radio and said goodbye to his wife live, and that of a desperate man who cursed his alleged executioners with tears in his eyes. Probably none of these are true, but if anything, one would lean toward the first. Komarov was a daring fighter pilot, used to facing death.

No one at Baikonur knew anything about the crash because, as usual, communication with the spacecraft was cut off during the descent. But when the rescue units arrived at the landing site and reported that "*the cosmonaut needed urgent medical attention*", it was understood that something serious had happened. The local aviation cut off the transmissions to avoid leaks of news. But this time there was no way to hide the tragedy from the world. Shortly after takeoff, Tass had issued a terse press release announcing the launch of a new type of probe, without mentioning the entire program. But Western analysts had sensed that it must be something big,[10] so they were all ears.

Defense Minister Ustinov was informed immediately, and an hour later Brezhnev, who was visiting Czechoslovakia.[11] In the evening the tragic news became public. It fell to Gagarin to remove the charred fragments of his

---

[10] Analysts had noticed that the name of the spacecraft was accompanied by an order number, *Soyuz-1*. Something unusual for a single prototype. This conclusion was in line with a note from Reuters that appeared four days before the launch. It speculated about possible Soviet wonders involving the docking of two orbiting spacecraft for some mysterious tandem adventure.

[11] It is said that the KGB officer who brought Gagarin's petition to Brezhnev was sent to Siberia to destroy the evidence that incriminated the premier.

friend's body from the wreckage. Komarov, the first official victim[12] of space-flight, left behind a wife and two daughters. His ashes, like those of Korolev, were buried on the wall of the Kremlin in eternal remembrance. Shortly thereafter, the same wall would welcome the mortal remains of another Soviet space hero, Yuri Gagarin himself.

---

[12] There are rumors of other victims among the Soviet cosmonauts, but no official or even unofficial information has ever been released.

Fly Me to the Moon

*Now is the time to take longer steps – the time for a great new American enterprise
– the time for this nation to take a clear leading role in the conquest of space which, in
many ways, could represent the key to our future on Earth.*
John F. Kennedy
*There are only two problems to solve when going to the Moon: first, how to get there;
and second, how to get back. The key is not to leave before you have solved both.*
Neil Armstrong.

On Wednesday morning, March 27, 1968, at 10:15, a single-engine jet MiG-15UTI[1] with Gagarin and his instructor, Colonel Vladimir Seryogin, a pilot of proven experience, took off from the Chkalovsky airfield near Shchyolkovo, Moscow *Oblast*. Low clouds covered the sky over the snowy birch forests surrounding the capital. In response to the tragic loss of Komarov, Yuri had resumed his former activity as an aviator. His friend Vlado was retraining him to tame the magnificent single-engine jet designed in Mikoyan's OKB-155, a nervous and quirky thoroughbred. The mission should be very short. A quarter of an hour after takeoff the two had already completed their assigned program, and Gagarin asked the tower for permission to return. From that moment the radio communication was interrupted.

---

[1] MiG, a familiar acronym from the 1950s, stands for Mikoyan-Gurevich, the names of the two designers of the Red Army jet aircraft.

Three hours of frantic search. Finally, the jet was identified by its remains among the trees of a dense forest near the town of Kirzhach, in the Vladimir *Oblast*. The bodies of the cosmonaut and his companion lay lifeless among the still smoking wreckage. After the shock there was a great mourning in Chkalovsky, in the Soviet Union, and in the whole world. The immortal Icarus, the peasant turned eagle who had given mankind the key to the sky, had died. The KGB, the government, and military aviation each launched an independent investigation to determine the causes of the tragedy, but according to established Soviet practice, the conclusions were not made public. It was only said that Gagarin had been the victim of an accident, as in the first notice of Tass: "*It is officially announced here that the hero of the Soviet Union Yuri Gagarin, the world's first cosmonaut, has died in an air crash*". But what kind of accident?

In the absence of facts, insinuations and slanders flourished, including the most ridiculous ones: a political assassination ordered by the Kremlin out of jealousy or to eliminate an embarrassing myth; the result of drunkenness; or even alien abduction. More realistically, it was argued that the two experienced pilots had lost control of the plane to avoid a sudden obstacle, a bird or a balloon; or that they had failed to recall the aircraft after a sharp dive initiated at 3000 m to compensate for a loss of cabin pressurization,[2] perhaps because they had been badly informed at the morning briefing about the height of the last layer of clouds. Cosmonaut Leonov, who was flying nearby in a helicopter, later wrote that he heard two bangs. He concluded that Gagarin's jet had stalled due to turbulence caused by the supersonic speed of another plane, an SU-15 interceptor, which was not supposed to fly so low.

The 34-year-old space conqueror was given a state funeral and his ashes joined those of his "father" Korolev and his friend Komarov on the Kremlin wall in Red Square, near Lenin's tomb. An open-air memorial where the Russian people still revere their heroes. The grateful homeland would also remember him with a monument built for the 1980 Olympic Games in Moscow: a titanium statue over forty meters tall, placed in the center of Gagarin Square, along Leninsky Prospekt, one of the main arteries of the Russian capital. On the base an inscription: "*On April 12, 1961, the Soviet spacecraft Vostok with a man on board flew around the world. The first person to penetrate into space is a citizen of the Union of Soviet Socialist Republics, Yuri Gagarin*". A certificate of a great victory but also, if you know how to read it, an *excusatio non petita* of an equally resounding defeat in the race to the Moon.

---

[2] Someone insinuated that ground technicians had accidentally left a cockpit vent open while preparing the plane.

Difficult times were upon the world. Five days after Gagarin's funeral, Reverend Martin Luther King, leader of the African-American civil rights movement, was shot and killed in Memphis, Tennessee. It was another blow to President Johnson's policy of racial integration after the assassination of Malcolm X in New York in 1965. Five weeks later, 800,000 people marched in Paris in a massive left-wing demonstration that followed the incidents a few days earlier in the Latin Quarter between police and university students. These were the harbingers of the great movement of socially heterogeneous mass protest known as the French May and Sixty-Eight. And on June 5, just after midnight, the candidate chosen by the Democrats to succeed Johnson, Robert Kennedy, John Fitzgerald's brother, was assassinated in the Ambassador Hotel in Los Angeles by the revolver of a Syrian-Palestinian. To avenge, declared the murderer, the American support to the State of Israel in the Six-Day War, mother of many of the subsequent and serious problems on the Middle Eastern chessboard.

Only a few years earlier, in August 1963, Martin Luther King had said at the March on Washington for Jobs and Freedom:

I have a dream that one day this nation will rise up and live out the true meaning of its creed: "We hold these truths to be self-evident, that all men are created equal".

This crusade for civil rights polarized American society, where racism was still deeply rooted. But all, or most Americans, agreed on the national effort to be the first to conquer the Moon, as John Kennedy had called for and promised. This collective dream was realized thanks to the Gemini and Apollo programs, which not only won the race against the Soviets, but also demonstrated the technological and managerial superiority of the American model. This is a story whose prelude goes back several years.

The Gemini program was conceived as an evolution of the Mercury project, so much so that it was originally christened Mercury Mark II. Although launched after the Apollo program, it was completed before its older brother began its glorious ascent to the Moon. Twelve missions, the first two unmanned and the rest with a pair of astronauts—hence the name, which alludes to the mythological twins (*"gemini"* in Latin) Cartor and Pollux—in a Rossinian crescendo of brilliant successes similar, *mutatis mutandis*, to those first stages of the space race systematically won by the Soviets.

The goal was to test equipment, personnel, and procedures both in flight and on the ground to gain valuable experience that could be applied to lunar missions. The program included launches into Earth's orbit, mostly low, with durations ranging from five hours to two weeks, to test trajectory changes, rendezvous and docking with other spacecraft, extravehicular activities (EVAs), and targeted splashdowns, with astronauts in control of the maneuvers: real pilots and not simple passengers, guinea pigs, and testimonials to the perfect functioning of the machines.

Exactly the opposite of Korolev's strategy, which had been motivated by insufficient economic resources and the fear that the physical inadequacy of the human body to withstand the extreme conditions of space could compromise flight safety. The chief designer had started his triumphal adventure when there seemed to be no other option. Then, he had grown fond of it, with the result that extreme robotization deprived the Soviets of the added value of human intelligence in dealing with both ordinary and, most importantly, emergency situations.

The team of "Magnificent Seven" astronauts on the Mercury missions was strengthened by quadrupling the staff with new recruitment rules. NASA needed a broader and more articulate range of skills than those of the first heroic flights, when one brave fighter pilot could do the job. Twenty young men were selected, all in good health and college educated, most of them members of the armed forces or at least with a military background, but also one true civilian candidate. He will be the first human to set foot on the lunar surface. For each flight, three pairs were formed: an official crew, a reserve, and two astronauts to coordinate communications from Earth in centers in Houston and Cape Canaveral.

The Gemini capsule had the shape of an ice cream cone, truncated at the top and divided into two distinct and interconnected modules. The lower, more squat one contained the retro-rockets, fuel, and instruments. In front of it, a re-entry module with a pressurized cabin of just over 2 cubic meters, capable of accommodating two astronauts on ejection seats (an option particularly valuable at launch in case of an emergency). It had two small portholes, one for each seat, the spacecraft controls, the instrument panels, and the life support systems for the passengers. The bottom was protected by the heat shield for the final plunge into the atmosphere. At the top, a small cylinder housed the equipment for possible docking with another spacecraft and the three parachutes to slow down the descent, with diameters ranging from 2.4 to 25.6 m, to be opened sequentially from an altitude of 15,000 m.

The first test launch took place on April 8, 1964, from Cape Canaveral, now renamed Cape Kennedy,[3] to verify the durability of the Gemini spacecraft. The capsule was lifted into the sky by a two-stage Titan II rocket, the evolution of an ICBM manufactured by the Glenn L. Martin Company,

---

[3] The proposal came from President Johnson in a televised address on Thanksgiving Day 1963, just six days after the deadly Dallas attack, at the suggestion of John Fitzgerald's wife, Jacqueline.

which in 1961 had become the giant Martin Marietta Corporation.[4] A detail that helps us understand the importance of private companies in the space adventure of the United States.

After three orbits, the shuttle, still attached to the second stage, was left to its own devices. During its natural re-entry at the end of 64 orbits, it burned up due to the friction of the atmosphere as it flew over the Indian Ocean. To facilitate its demise in the air, four holes were made in the heat shield to weaken the structure. It was a precaution dictated by the lack of previous experience: safety had to be ensured to avoid the risk of the spacecraft crashing into a human settlement, and industrial secrets had to be protected to prevent the technology from falling into the wrong hands.

The complete success of this first mission convinced NASA officials that they could risk a manned flight. A crew was promptly assembled with a commander, Virgil Gus Grissom (1926–1967), and a pilot, John Young, along with a backup crew made up of Walter Schirra and Thomas Stafford (1930–2024). But on the scheduled launch date, November 16, 1964, nothing happened. Lightning in August was blamed for damaging the launch pad. An explanation that seems to be a half-truth. It is more likely that the program was stopped by a delay in the Gemini ground tests.

Then, there were the hurricanes, three of which came out of the Caribbean and hit Florida, creating weather conditions unsuitable for a launch. The severe damage to infrastructure in this case was a blessing in disguise, as it allowed NASA to justify delays to Congress, which was now keeping a close eye on space activities. Senators wanted victories, not talk, and had granted the space agency resources and freedom of movement usually reserved for generals at the front in wartime.

In this climate of strong psychological pressure, an unmanned suborbital ballistic flight mission was launched on January 19, 1965. The most important test concerned the strength of the heat shield, the astronauts' life preserver. A malfunction could have meant certain and horrible death, as the passengers would have been roasted by the heat of friction with the air in the re-entry module. NASA also wanted to test some innovative subsystems, such as fuel cells to replace solar panels. Devices that could convert chemical energy directly into electricity. An improvement the Soviets had not yet considered.

Gemini 2 had climbed to an altitude of 170 km; then, having exhausted its momentum, it had meekly surrendered to the arms of gravity, which always demands payment of its claims. At the end of a parabolic trajectory similar

---

[4] In 1995, it will become Lockheed Martin Corporation after the merger with the powerful Californian aerospace company.

to that of a stone tossed into the air by a muscular slingshot, the lifeless shell had plunged with appropriate grace into the Atlantic Ocean, 3400 km from Cape Kennedy. Eighteen minutes of flight in all, with everything working perfectly except the fuel cells.

Finally, on March 23, 1965, the crew that had been waiting on the ground, formed by Grissom and Young, was launched on Gemini 3 for an orbital flight of almost 5 h, ending with a splashdown near Grand Turk Island, a territory of Her Majesty Queen Elizabeth in the North Atlantic. The primary purpose was to qualify the astronauts in various maneuvers and to prove the shuttle's guidance system from any place on Earth: a set of facilities that the Soviets did not fully possess. The test also included the functionality of the Mission Control Centers at Cape Kennedy and especially in Houston.

After reaching the parking altitude in a few minutes, Grissom took over the controls of Gemini and began to steer and change the trajectory. These were not just simple attitude adjustments, which the Soviets had already exper-imented with extensively, but true trajectory changes. A bit like steering a boat pushed by the wind. Grissom and Young had learned these maneuvers in a simulator in St. Louis, Missouri, at the headquarters of the shuttle manu-facturer, the McDonnell Aircraft Corporation, and then at Cape Canaveral. In addition to other routine training, of course, such as flying fighter planes, parachuting, prolonged centrifuge sessions, and getting used to both weightlessness and exiting a capsule floating on water in a pool.

After the third orbit, the two astronauts descended to a very low altitude, only 84 km, to be able to resort to a quick natural re-entry in case the retro-rockets failed. The story goes that flight controllers at Cape Kennedy were furious because Young, being a practical joker, had smuggled a sandwich aboard the Gemini. This was a serious violation of protocol, which prohibited any form of food other than toothpaste tubes to prevent dangerous crumbs from floating around the cabin. Perhaps this is how it happened. "*What stuff is that?*", Gus had asked John. "*A rye bread and jerky sandwich*", the latter replied. "*Where the hell did it come from?*". "*I got it from 'Wally' [Schirra] and hid it in my spacesuit. Let's see what it tastes like. You can smell it, right?*". A dialogue worthy of two cowboys around a campfire in a film about the glorious epic of another conquest, that of the Far West. Hard to believe, but appreciated by a certain America that has not yet shed the clichés of a national history that is too short. In any case, this funny episode was another useful test for NASA to update the way astronauts are fed in flight.

The descent went basically well, except for the maneuver where the shuttle had to be inverted at the change from the second to the third parachute in order to adjust it for the entry into the water. A too abrupt execution caused the two astronauts to hit their heads against the porthole so hard that Grissom's helmet visor broke. The splashdown happened far away from the expected location. The astronauts remained in their well-sealed capsule for half an hour, fearing it might sink and even suffering from seasickness. The height of irony for those who had endured tremendous accelerations and floated in weightlessness. Then, they were rescued and brought to safety. All in all, a success. Now it was possible to take a leap forward and try to catch up with the fleeing Soviets.

The real comeback happened with the launch of Gemini 4 from Cape Kennedy on June 3, 1965. On board were two rookies, James McDivitt (1929–2022) and Edward White (1930–1967), who orbited the Earth 62 times in four days. Much better than Phileas Fogg and his valet Passepartout, Jules Verne's characters, who a century earlier had taken eighty days to circumnavigate the globe. In addition to various scientific experiments and some technological checks, the mission included two novelties. An attempt to rendezvous with the second stage of the Titan II vector and an extravehicular activity, now made less exciting by Leonov's feat two months earlier. "*Only out there can you feel the size, the immensity of all that surrounds us*", the Russian cosmonaut had said. White would also experience that exhilarating feeling, but he was only second. "*Lord, give my enemy strength and let him live long so he can see my triumph*", Napoleon Bonaparte rightly claimed, because "*he who fights can lose, he who doesn't fight has already lost*", Ernesto Che Guevara added. It was time for Americans to taste the joy of being first.

The Gemini approach to the Titan stump failed due to the rapid depletion of propellant available for the maneuver. Instead, the EVA was successfully completed for a total of 23 min, twice as long as Leonov had been floating in space, and without the need to keep the cockpit pressurized to ensure cooling of the electronics. A step forward in simplifying procedures by eliminating the depressurization chamber. White showed great adaptability to the vacuum. He used a compressed air pistol to assist his movements, which soon ran out of air. A kind of mini rocket, it was very risky to operate, as the slightest mistake could have shot the astronaut far into space, making his return problematic if not for an 8-m umbilical tether anchoring him to the spacecraft. A seemingly obvious precaution destined to cause some problems. In the absence of gravity, the long tether seemed to have a life of its own, tangling itself in every possible handhold and hampering the spacewalker: a snake gone mad in convulsions. But "*it is fun*", White answered when he was told to end his spacewalk. "*I come back in, and it's the saddest moment of my life*". A boast? Probably, but also a sign of a growing familiarity with space.

Now master of Earth, water, and air, mankind gradually took possession of empty space. Empedocles would gloat.

After almost one hundred hours of flight, Gemini 4 took the sea route, missing the planned splashdown point by a good 80 km. This was caused by a malfunction in the onboard computer, which forced McDivitt to perform a manual splashdown. The astronauts were recovered safely. The record for staying in orbit remained in Soviet hands thanks to Bykovsky and his flight of almost five days. But now the chasers were within a hair's breadth of the leaders, showing a unity of purpose and ability to plan that the competitors behind the Iron Curtain had not even dreamed of. Korolev realized that the game had started again and victory by no means certain, so he looked for another ace up his sleeve to close the gap. Unfortunately, he had played all four and the Americans were attacking him from all directions.

On July 14, 1965, Bastille Day, the Yankees had the satisfaction of showing the world 22 photos of the planet Mars, taken by the Mariner 4 probe launched on November 28, 1964, scanned in situ and then sent to Earth in digital form. The signals had arrived from a distance of half a billion kilometers, an absolute record for the waves discovered by Guglielmo Marconi. An important scientific and technological achievement, but not enough to put the American taxpayer in a good mood.

So came Gemini 5, whose launch was delayed from August 19 to August 21, 1965, due to a technical malfunction and bad weather. On board were Gordon Cooper and Charles Pete Conrad (1930–1999) to experiment with a long cruise and the prolonged absence of gravity, along with a simulation of a rendezvous and the associated equipment and procedures. As sparring partners, they would have used a facsimile of the flange, which they would have carried into space and then released into orbit.

Despite a fuel cell problem that nearly scrubbed the mission, several rendezvous attempts were made by piloting the spacecraft until the guidance system failed. The arrival of a tropical typhoon in the splashdown area prompted mission control to order an early return after 120 orbits and 191 h in space. The astronauts were recovered by air-lifted divers 150 km from the planned landing site and brought to safety aboard a Navy ship. Another success. Now, the duration record was in the hands of the United States. On the scoreboard of the cosmic billiard, the white and blue dots were dangerously close to the red ones.

The Gemini program moved forward like a steamroller, aided by good fortune. For example, Gemini 6 could not be launched as programmed because the sparring partner spacecraft with which to train for rendezvous and docking was unavailable. This was not a simple flange, but a highly techno-logical Gemini-Agena Target Vehicle (GATV), consisting of the second stage of a Lockheed Agena rocket, equipped with a docking system manufactured by McDonnell Aircraft Corporation (the one of the famous DC9). Due to a problem with the launch vehicle, the GATV-6 had failed to enter orbit and NASA had destroyed it in flight.

Without a partner, the Gemini 6 mission became useless and was aborted at the last minute with the crew already on board. A great disappointment for NASA and a great irritation for the whole country. But someone came up with a bold proposal. Send Gemini 7 into orbit as well and make the rendezvous directly between two spacecraft, cutting out the middleman. President Johnson liked the idea so much that he announced it to the nation on television. It was a risk—no one knew if it was possible to launch two spacecraft within half a month from the only available launch pad—but once the White House endorsed it, it had to be done. So James Webb decided to put his face and neck on the line and invade Soviet territory. In fact, the Reds had long cherished the idea of building a space station in low Earth orbit by assembling modules the way children do with Legos and sending them into space in a procession of rockets.

To this end, Gemini 7 was launched on December 4, 1965, to stay in space for a full 14 days, a record that would remain unbeaten until the flight[5] of *Sojuz-9* in 1970. The two astronauts, Frank Borman (1928–2023) and James Lovell (1928–), were to prove that they could withstand long periods of weightlessness by experimenting with lightweight suits, very comfortable but inapt for survival in the event of rapid depressurization. They adopted the work rhythm of Earth days to conduct scientific and medical experiments. The plan of alternating rest shifts had in fact failed miserably in the face of the impossibility of a minimum of privacy for sleep. While waiting for Gemini 6, they also used their time to prepare for rendezvous with the second stage of Titan II, which was oscillating dangerously. Gemini 7 could not be brought closer than 15 m to the target.

On December 15, 1965, after a first attempt was aborted without major consequences thanks to the cold-bloodedness of the crew, the Gemini 6 mission was launched with Walter Schirra and Thomas Stafford aboard. There was great concern about the ground management of two simultaneous missions. But all went well. The two spacecraft cleared an initial separation of 1900 km (the straight-line distance from Washington DC to Dallas TX) in just over five hours, then approached each other at a relative speed of only 3

---

[5] A 17-day and 17-h mission, aimed at verifying human resistance to long exposure to zero gravity.

cm/s (100 m per hour, like some August freeway jams). This was nothing new. In fact, three years earlier, the *Vostok* spacecraft had come within a small distance of each other, but automatically, through precise calculations of launch trajectories. Instead, the two American spacecraft had done it by taking advantage of their agility of movement and the pilots' ability to guide them to meet.

In fact, all the work fell on the shoulders of Gemini 6, because the other spacecraft had already emptied its fuel tanks during its long stay in orbit. At the moment of closest approach, at the mutual distance of 40 m, it is said that the pilot of Gemini 7 asked his companions on the other spacecraft: "*How's the visibility?*", to which Schirra replied: "*Pretty bad, I can look out the window and see you guys inside*". The average American was amused by these jokes laden with innocent bravado. Much more serious was an earlier comment "Wally" had made about terrestrial pollution during a conversation with an Associated Press journalist: "*I was very impressed by the fact that when I flew over India and China - this was in 1962 - both countries were very, very extensively covered with clouds of dust and smoke, whereas Africa was quite clear, which proved to me that even then something was going wrong with the environment*".

The two shuttles remained 110 m apart for a while, playing blind man's bluff. Then, after five hours of back and forth, they moved away from each other to allow the astronauts to sleep peacefully. It is almost like watching two galleons at anchor in an adventure movie, separating in the evening to drop anchor and send the crews to sleep without the risk of accidental collision.

At the 15th orbit, after a 25-h flight, Gemini 6 was brought back to Earth. The splashdown was broadcast live on television: far more powerful message than any Tass press release. It was necessary to repay the millions of citizens whose taxes had contributed significantly to the realization of these ventures. Gemini 7 returned with 330 h of uninterrupted navigation. The astronauts were all well; in fact, "*better than expected*".

Then, on March 16, 1966, Gemini 8 arrived with Neil Armstrong, an aeronautical engineer and test pilot, and David Scott,[6] a Texan fighter pilot, son of art. They had the honor of crowning a rendezvous with a full docking to a GATV launched a few hours earlier. It happened at the fifth orbit. Everything was going well, but as soon as Armstrong attempted a simple maneuver, the coupling between the two vehicles became unstable. Startled, the two astronauts immediately released the partner, but strangely, the oscillations got

---

[6] At the end of the Apollo 15 mission in 1971, David Scott performed Galilei's famous experiment on falling graves, demonstrating that on the Moon's soil, in the absence of air, the fall time is the same for a feather and a hammer.

worse. The Gemini 8 spun around like a crazy top. Scott did not lose control and somehow managed to stabilize the shuttle before the whirling rotation overpowered him. Danger averted. What had happened? One of the nozzles in the attitude correction system had failed to close, causing a constant twist. But still nobody knew. The mission was immediately aborted after only seven orbits because the stability recovery maneuver had consumed too much fuel. The GATV was left in orbit for future use.

The splashdown took place 800 km from Okinawa, the Western Pacific Island made famous by the battle between U.S. Marines and the Imperial Army of Japan. Immediately rescued by divers launched from an aircraft, the two astronauts were picked up three hours later by a Navy ship. Too much time and another lesson learned. The crew recovery procedure had to be improved. Do it faster and do it better. The same rules that govern Formula 1 team technicians changing tires during a race. Step by step, NASA fine-tuned the ingredients to make the long journey to Earth's satellite a safe one.

It was now clear that the Lunar Championship had a new leader: a solid, determined team, rich in resources and endowed with ingenuity, imagination and method. Obviously, the Soviets had not yet accepted their new role as spectators. As the Italian poet Giacomo Leopardi once wrote, "*The most solid pleasure in this life is the vain pleasure of illusions*". They lived in the usual hope of a secret weapon, a super-powerful rocket that could change the course of history. But Korolev's *N1* rocket never took off. It was too big, too complex, too expensive, and it ended up in hands that were not up to the task. Anyway, the Reds still managed to land three well-placed blows, then spoiled by the fatal Komarov's accident that exposed them to the vicious disparaging propaganda of Westerners accusing them of being careless with the lives of their cosmonauts.

On February 3, 1967, the Luna 9 probe, although unmanned, arrived safely on the surface of the Earth's satellite at the end of a 79-h journey. It had been launched from Baikonur at dawn on January 31. The first three stages of the standard *Molniya* rocket carried a train consisting of the fourth stage of the rocket and the spacecraft itself into low Earth orbit. With the extra thrust of the engine still available, Luna 9 was placed in a very elongated geocentric orbit with an apogee of 500,000 km. A few course corrections, a targeted braking, and a complicated series of maneuvers had led to the impact with the Moon. The capsule containing the instruments was jettisoned from a height of only five meters to glide at very low speed onto the lunar surface. Two or three jumps and … voilà! The first intact human artifact, with working eyes and ears, lay undamaged on the sand of *Oceanus Procellarum*, ready to transmit images collected by a camera. Another record for the Russians and another gift from Korolev, since the probe had been designed by his OKB-1 during his lifetime and then entrusted to another bureau.

The joy of the people of the USSR was renewed two months later with the arrival of *Luna 10* in a selenocentric orbit: a satellite of the satellite, which at every turn ideally saw another piece of Soviet technology resting on the lunar dust. It had never happened before and demonstrated the indomitable vitality of space engineers from behind the Iron Curtain. Only the footprint of a human foot was missing, but contrary to the adage that there is no two without three, this "*small step*" would not be taken by a red foot with a sickle and a hammer tattooed on it.

In the middle of these two lunar missions, on the first of March, the interplanetary ship *Venera 3* arrived on the blue planet. A ton of instruments that had been traveling in interplanetary space for three and a half months and that, according to the intention of the designers of OKB-1, should have been gently set down on the Venusian soil. However, due to one of the usual failures, the spacecraft had instead plunged headlong and unbraked into the thick mantle of clouds enveloping the planet and crashed to the ground. The first landing, albeit catastrophic, of a man-made artifact on a planet in the Solar System other than Earth. The Yankees had tried to compete with Mariner 4 and 6, and with Ranger 6 and 7, but all the cups up for grabs had been won by the Communist engineers.

Meanwhile, the Gemini program continued methodically along its chosen path, making the most of the experience gained from time to time. On June 3, 1966, Tom Stafford and Gene Cernan (1934–2017)—a backup crew to replace the official one that had died in a plane crash during a transfer within the United States—were launched into Earth orbit for further tests on docking and spacewalks. They had been delayed due to the failed launch of both the sparring partner used to practice space hugging and a simplified GATV model. The latter had managed to reach orbit, but had brought along a protective shell covering the docking flange. So no docking, but just a rendezvous at the conservative distance of 8 m.

On June 5, Cernan made a walk of more than two hours in space, carrying on his shoulders a backpack the size of a suitcase, equipped with survival, communication, and movement systems, a product of the Air Force laboratories. The results were not exciting, though. In addition, Cernan had great difficulty performing the tasks assigned to him because his hands were constantly wandering in search of handles.

It is true that the usual umbilical cord secured him to the ship. But the slightest push was enough to kick him away from his workstation or make him do a somersault. Sweat, poorly controlled by the helmet's air conditioning system, fogged up the visor, causing further anxiety for the spacewalker. Now NASA knew that something had to be done. After three days and 45 orbits, Gemini 9 was called back to Earth and splashed into the waters of the Western Atlantic Ocean just 700 m from the expected impact point. The docking program was not complete, nor were the activities planned for the very long spacewalk. But Leonov's record was broken!

A month and a half passed and it was Gemini 10's turn. John Young and Michael Collins (1930–2021) had the task of testing the usual maneuvers: a rendezvous and docking with a GATV, launched a few hours before and successfully this time, and two extravehicular activities, along with many scientific and medical experiments. The docking was performed on the fourth orbit, after having followed the target for 1600 km, but with a fuel consumption twice higher than expected. For this reason, Houston asked the astronauts not to separate from the target vehicle in order to use its engines for maneuvers. In fact, thanks to the GATV engines, the spacecraft was lifted to an altitude of 760 km. A new record, as *Voskhod-2* had reached an apogee of only 475 km. More and more USSR flags were lowered and replaced by American ones.

Having freed itself from its own GATV, Gemini 10, like an unquenchable golden retriever, went on the hunt for the target vehicle left in orbit by Gemini 8 and, having located it, came within three meters of it. Collins, who had already tasted the void by looking out of the spacecraft without completely leaving it (stand-up EVA), left his safe nest to reach the target by walking in space. He had only 25 min of autonomy to retrieve a micrometeorite detector attached to GATV-8, which had been exposed for several months and should have been loaded with findings. He succeeded, and that was what mattered to the lunar mission strategists. However, during the difficult re-entry inside the Gemini, he lost both the detector and the camera, much to the disappointment of the scientists.

Some mistakes were made. However, there was a growing sense on both sides of the Iron Curtain that the Americans were now the masters of orbital space. In particular, the crucial operation of docking two separate units in orbit had finally been achieved, thanks to the intelligence of automatic systems and also—but the Soviets had not yet fully realized this—thanks to the assistance of skilled pilots. On the 43rd orbit, Gemini 10 was brought back with a splashdown in the Western Atlantic, 875 km from Cape Kennedy and only 5 km from the expected landing site.

Gemini 11 flew on September 12, 1966, with Pete Conrad and Richard Gordon (1929–2017) on a three-day mission. Once again the program consisted of a docking test with a GATV previously launched into orbit. It worked on the first attempt, with even less fuel burn than expected. Driving school paid off. Each of the two astronauts repeated the exercise twice to consolidate the procedure. Then, they fell asleep while remaining connected to the target. Gordon's EVA was scheduled for the following day and lasted 107 min. The spacewalker had to take one end of a 30-m tether attached to the GATV and connect it to Gemini, but the effort of the operation darkened his view so he quickly returned to the spacecraft after only 26 min of exposure to space. The next day the GATV engine lifted Gemini to an altitude of 1370 km, a new record for distance from Earth, surpassed only by Apollo 8. Back to low altitude, Gordon leaned out of the shuttle to take photos during a two-hour stand-up EVA.

When the cabin could be pressurized again, the two astronauts performed a unique maneuver. A sort of tug-of-war and cartwheel. They extended the 30-m tether connecting the Gemini to the GATV and rotated the two objects around their common center of gravity. A feat that demonstrated absolute mastery of the vehicle and the procedures. After taming, like a Native American with his Mustang, the inevitable oscillations caused by the improbable configuration, Pete and Richard enjoyed for the first time a touch of artificial gravity recreated in orbit. Nothing their bodies could feel, but it was noticeable in the objects floating in the cabin air. Then, they released the tether and performed another rendezvous *ad abundantiam* (beyond the need), without even using the radar, which had broken down by then. On the 44th orbit, they fired the retro-rockets and descended, for the first time fully automatically; something the Soviets almost always did. A quiet splashdown in the Western Atlantic ended the 71-h flight. Mission accomplished.

Two months later, on November 11, 1966, the launch of Gemini 12 completed the program. Commander Jim Lovell and pilot Buzz Aldrin (1930–) had the task of summarizing all the lessons learned in three years of tests on approach and docking techniques. And indeed, during the third orbit, they easily docked a GATV launched by an Atlas rocket. They had to maneuver by sight because the radar was broken again. Another first. The procedure was repeated, but they had to abandon the use of the target vehicle's engine for fear of a malfunction. There was a risk of being shot out of orbit, far from Earth. The hazards were no longer necessary, and, above all, they were especially unwelcome to NASA, which feared the impact of disasters on its budget.

The remaining problem was how to spend the time during the 59 orbits. The two astronauts decided to observe a total solar eclipse that crossed South America, and Aldrin took a long time to analyze his reactions in a stand-up

EVA that ended with the first selfie in space history. The next day he went out again, this time for a real walk, visiting the target vehicle still attached to Gemini 12 and even cleaning the spacecraft's portholes: an unlikely glass washer who confirmed the versatility of American university students. In fact, Aldrin was the only astronaut in the group with a Ph.D.[7] The next day he performed yet another stand-up EVA. Then, with some trepidation, the spacecraft returned to Earth, 5 km from the planned location. The last act of a well-organized program that had given the Americans full control of the procedures for managing activities in orbit. The opponent in the red jersey, who had frightened them at the beginning of the stage with a sudden and unexpected escape, had been caught and overtaken and was now pedaling with difficulty but still not tamed, even hinting at a sprint.

It was the right moment to use the difficulties of the opponents and the quarrels among the various cockerels of the rich and diverse Soviet archipelago to deliver the coup de grace. The weapon was ready. It was called the Apollo Program, in the making since 1961, when Kennedy announced the crusade to "*land a man on the Moon and return him safely to the Earth*". Its predecessors, Mercury and Gemini, had been successful. Now, it was time to make good on the president's promise. Since a powerful and reliable launcher was needed, it was decided to pragmatically enlist the help of the great rocket pioneer, Wernher von Braun. Leaving the familiar low Earth orbits to venture into deep space, at distances immeasurable on a human scale, the explorers of this new frontier had to be assured of a reasonable chance of survival.

To the ordinary American taxpayer, the economic cost of the operation might seem exorbitant compared to the immediate impact on his daily life, except for the satisfaction of teaching the Bolsheviks a lesson. But to the shrewd and efficient private sector, it was clear that the conquest of the Moon would stimulate unimaginable advances in many fields of science and technology, including avionics, computers, materials, and telecommunications, with enormous returns to the wealth and prestige of the country. Pride, money and prosperity, then, versus patriotism and a utopian and heroic celebration of socialism, a fig leaf for other political designs and less noble ambitions. James Webb was well aware of this and would reiterate it in 1966, at a hearing of the U.S. Senate Committee on Aeronautical and Space Sciences, Washington, D.C.:

I think that lunar landing is mistakenly thought to be the major goal. The capability to operate in space, to see what you can do and what you cannot do, and

---

[7] In 1963 Aldrin had received a doctorate in astronautics from MIT, defending a thesis on some aspects of manned orbital rendezvous.

then proceed to do the things that you can do that are useful to the nation is going to come out in the 1968 budget.

Once the goal was set, the overall strategy had to be determined. There were two options. Fly directly to the Moon with a yet to be developed super rocket, or reach the Moon after a technical stop in a service area. Wernher von Braun, for example, proposed to halve the loads to be launched by separating the spacecraft from the fuel tank: a solution that Korolev had also considered. The two units would have been launched separately into Earth's orbit and then joined together to go to the Moon, make the landing, and then return with the help of Earth's gravity, using everything that had already been tested by the Gemini missions. It seemed like the Columbus egg.

But there was a group of NASA engineers who thought otherwise. Their spokesman was John Houbolt (1919–2014), a 40-year-old man from Iowa with a Ph.D. from the prestigious Zurich Institute of Technology. His proposal turned the perspective on its head. The service orbit would have to be around the Moon, and instead of assembling something, the spacecraft would have to be split into a Command Service Module (CSM), which would remain in orbit and be used to return to Earth, and a lightweight Lunar Excursion Module (LEM), which would be used as an elevator to descend to the satellite and then climb back up to catch the bus home.

The Lunar Orbit Rendezvous (LOR) increased the risks. If something went wrong, gravity, which in low Earth orbits ensured a natural return, could not be relied upon. Houbolt was accused of giving the astronauts *"a 50 percent chance of getting to the Moon and a 1 percent of getting back"*, as he himself reported in a 1961 letter to NASA Associate Administrator Robert Seamans. However, his plan represented a shortcut that could put America back in the game when the Soviets were still leading the race. Pretty soon, even von Braun was persuaded to support it.

The idea was not new. Besides Oberth, it had been developed during the First World War by a Ukrainian scientist, Yuri Vasilievich Kondratyuk (1897–1942), a pioneer of astronautics and space travel, a theorist and a visionary. A character with a life on the edge of a fairy tale, worth telling also because his role in the lunar saga was so significant that it was officially recognized by NASA and by the same Neil Armstrong.

Born into a wealthy family in Poltava, then a city of the Russian Empire in southeastern Ukraine, Kondratyuk was named Aleksandr Ignatyevich Shargei. His father had studied physics at the university. His mother, a French teacher and civil rights activist, came from a line of famous military men. Declared insane, she was sent to an asylum, and Aleksandr, still a child, was entrusted to his paternal grandparents. While he was studying at the Great

Polytechnic in Petrograd, the Tsar declared war on the Austrian Emperor and Aleksandr was conscripted to fight against the Black Eagles. Sent to the North Caucasus to command a machine-gun platoon, he found time to complete notebooks of thoughts and calculations about space travel that he had begun at the age of seventeen. Another Tsiolkovsky with an even more unique destiny than the Kaluga genius. The concepts anticipated in the notebooks, found by chance in the oil cup of an engine, concerned fuel cells powered by solar heat, a multi-stage rocket to go to the Moon using a parking orbit around the satellite, a space suit, the attitude control of rockets by a gyroscope, and the famous gravitational slingshot to gain thrust from a "well-tempered" encounter with a celestial body.

In 1917, when the armistice with the Central Powers was signed, he tried to return home. Forcibly conscripted by General Kornilov's Whites,[8] he deserted and fled to Poltava; but his name was now on the blacklist of tsarist officers and deserters. He then tried to emigrate to Poland, without success. Blocked by the Bolsheviks, he risked execution and was released only because he was seriously ill with typhus. There was no point in wasting a bullet on such a human wreck. Instead he was able to heal. With the Damocles sword of defection hanging over his head, he decided to change his identity to that of a dead man and leave the places where he could be recognized. Under the new name of Yuri Kondratyuk, he settled in Novosibirsk, Siberia. Here, after organizing his notes into a neat text, he tried to publish them in Moscow. After some back and forth, and despite an excellent review from Moscow scientists, in 1929 he had to resort to a local printer. He paid him out of his own pocket with the little money he earned as a mechanic, and he even had to help the typographer compose the abstruse mathematical formulas contained in the meager 72 pages of the text entitled *The Conquest of Interplanetary Spaces*.

Difficult times were raging in the Russian Soviet Federative Socialist Republic, a huge country, fresh from a revolution and a bloody civil war that lasted more than five years. Anyone could be accused, rightly or wrongly, of being an "enemy of the people", with devastating consequences. All it took was a simple misstep. Kondratyuk's misstep was a consequence of his multifaceted genius.

He had designed a huge wooden silo to store grain, eliminating the need for nails, which were in short supply in Siberia. This measure was interpreted by the NKVD as an attempt to weaken the structure. Sabotage, which had to be punished accordingly. Sentenced to three years in a *gulag*, Yuri was

---

[8] The White Movement was a free political association with a military arm and anti-Communist aims, established in 1917 and supported by Western nations.

soon transferred to a *sharashka*, where he used his creativity to study alternative sources of rural energy. For good luck he was noticed by the People's Commissar for Heavy Industry, released, and sent to Kharkiv, southeast of Poltava. Passing through Moscow, he met the young Korolev, who tried in vain to capture him for his space projects. Who knows what would have happened if he had succeeded.

Kondratyuk spent the rest of his short life working on wind turbines. At the outbreak of World War II, he was drafted into the infantry and died under unknown circumstances on February 25, 1942, in the village of Krivtsov, Orel *Oblast*, without ever having met Tsiolkovsky and 27 years before his space speculations became reality. His name appears in the International Space Hall of Fame at the Museum of Space History, Alamogordo, New Mexico, next to those of Tsiolkovsky, Korolev, Gagarin, and Leonov, Oberth and von Braun, Armstrong, Aldrin and many American astronauts, Galileo Galilei and Isaac Newton. A great compatriot of his, Nikolai Gogol, had written: "*What a curious and attractive, yet also what an unreal, fascination the term 'highway' connotes! And how interesting for its own sake is a highway!*". And what highway is more magical than the one that leads to the beautiful Selene?

Let's now get back to the Apollo program. After much uncertainty, the mission architecture was approved in November 1962 and eleven aerospace companies were invited to bid. The LOR required a mega rocket to carry the astronauts into lunar orbit. Wernher von Braun himself and Arthur Rudolf (1906–1996), another talented Nazi scientist who had been transferred to the U.S. in 1945 as part of Operation Paperclip, took care of that. In the offices and laboratories of NASA's Marshall Space Flight Center, German words or entire speeches were often heard that now sounded like so many reassurances. Hitler's magicians had renounced the swastika and were working for the stars and stripes. The result of their efforts was the Saturn V, the largest rocket ever built by man: a giant 111 m high, 10 m wide, with a net mass of 130 tons that became 3000 tons after refueling. It consisted of three stages using liquid propellants, manufactured by Boeing, North American Aviation (the company that supplied the B-52s to the Air Force), and Douglas, respectively.

The launch of the Apollo program led to an enormous change in the size of the space agency, which had sole responsibility for the enterprise. In less than a lustrum, NASA's workforce quadrupled to 36,000 units. The systematic use of outside contracts, awarded to private companies but also to university institutions such as the Massachusetts Institute of Technology, provided jobs for nearly 400,000 people. Entirely new research and development centers were created. Starting in 1962, the Agency's total budget shot up to $5 billion a year, the equivalent of today's 34 billion euros, more than ten times what the Soviets were able to invest in space activities. Of this monumental sum, 60% was spent on Apollo, primarily to build the Lunar Module and Saturn V, to conduct meticulous ground tests, and to develop the computer systems needed for design and flight.

The requirements were stringent: powerful and re-ignitable engines, with self-igniting propellants to eliminate the risk of ignition failure, and above all, high reliability. NASA asked designers and contractors to certify an overall hardware failure probability of less than 1 percent and a crew loss probability of 0.1%. To achieve such an ambitious goal, the rule of redundancy was established. Wherever possible, every critical piece of equipment had to have either a backup or a Plan B for use in an emergency.

The first step in the strategy to reach the Moon and return was to select the trajectory and identify the launch window. The lunar landing had to be on the visible side of the satellite to maintain constant radio contact, and at the beginning of the lunar day (which is 29 Earth days) to allow the LEM pilot to take advantage of the long shadows in selecting the appropriate landing site. It was imperative that the Lunar Module land upright, otherwise it would be impossible to restart, resulting in the loss of the crew.

The 11 manned Apollo missions sent into space from October 1968 to December 1972 all had similar sequences, gradually becoming more and more complete one mission after another, until the final one. Each began at the end of the countdown. Ten seconds after ignition, the five engines of the Saturn V's first stage reached a thrust equal to the weight of the entire rocket, which could hold itself and lift off with the Apollo spacecraft on top. The explosive bolts anchoring the giant to the pad were then detonated and the amazing white tower took flight, burning 15 tons of fuel per second. In two and a half minutes, it reached a speed of 10,000 km/h at an altitude of 60 km. The first stage was then jettisoned and crashed into the ocean off Cape Kennedy.

Now, it was the turn of the second stage, seven times less powerful than the first because the mass to be accelerated had been significantly reduced. After tripling its altitude and speed, it too was jettisoned with its tanks empty to make way for the third and final stage. For now, its mission was to orbit the Earth with its payload: the CSM, a spacecraft similar to Gemini but ten times heavier and capable of carrying three astronauts, and the LEM, which looked more like a cubist painter's nightmare than a spacecraft. It did not need an aerodynamic shape because it had to operate in the absence of an atmosphere. The LEM itself consisted of two parts: a unit to descend to the Moon and a cabin connected to an engine to return to lunar orbit.

After one and a half revolutions around the Earth, the third stage reignited its engine to enter a transfer orbit to the Moon. With that task accomplished, the astronauts housed in the CSM went into action. They had to undock the Command and Service Module from the final stub of the Saturn, rotate it by 180°, hook the LEM to its tip, extract it from its container, and continue the journey to the Moon while the now useless third stage was lost in space. The crossing took a little over three days, during which a slow rotation of the Apollo around its axis of symmetry favored an even distribution of solar heat.

Once in the target area, with accuracy assured by a few minor course corrections directed by Mission Control in Houston, the CSM engine was fired long enough to slow the spacecraft and place it in a lunar orbit. The approach trajectory was designed so that if the engine had not worked, the Apollo spacecraft would have orbited the Moon and returned directly to Earth to be captured by the planet's gravity. A handful of revolutions around the Moon to assess the situation, then a controlled engine thrust converted the lunar orbit into an ellipse with a pericenter only 15 km from the surface of the satellite and an apocenter at 100 km. Heights unimaginably low for Earth satellites due to atmospheric friction. Now the LEM, to which two of the astronauts had transferred, could be released for the lunar landing. With zero tangential velocity, the LEM was taken over by the Moon's weak gravitational pull. A retro-rocket was sufficient to counteract the low force (parachutes could not be used due to the lack of an atmosphere).

The pilot, who was trained to handle the lander like a helicopter, guided the descent of the LEM on the best terrain with horizontal movements. All this in a short time and with little fuel. After touchdown, the operations on the surface of the satellite could take place: an EVA *sui generis* due to the presence of gravity, even if weak (making a 70 kg man feel as if he weighed only 12 kg). Then, after the return of the walkers inside the LEM, the upper part of the lander lifted off with its own engine, using the descent module as a launch pad. This was followed by entry into lunar orbit, rendezvous with the Command Module, docking and return to the CSM of the two astronauts who had landed on the Moon.

The mission was nearing completion. With the CSM in the opposite position to Earth, its engine was fired to resume the transfer trajectory (now to be retraced in the opposite direction). Another critical moment, as an engine failure could send the astronauts into an irreversible heliocentric orbit, condemning them to certain death. After that, three more days of travel and a real EVA to retrieve material from the Service Module (photos and micrometeorite traps). Then the arrival near the Earth, the jettisoning of the Service Module, and the re-entry of the Command Module alone, slowed down by the heat shield and the parachutes, until the splashdown in a little frequented marine area.

"*And if not now, why let your tears flow?*", asked Count Ugolino to Dante closing the account of his tragic story. Certainly, there is no need to cry when reading the program of this "excursion to the Moon", but there is enough to be afraid of. Too many things could go wrong, doom the mission and lead to the death of the crew. Kudos to those who, fifty years ago, dared to repeat that somersault a few times with surgical precision. From Apollo 7 to Apollo 11, everything went as well as a miracle or a fairy tale. The only exception was the mission that started the program, when astronauts Grissom, White and Roger Chaffee (1935–1967) burned up in the fire of the Apollo 1 spacecraft during a stationary test on January 27, 1967.

Grissom had said: "*If we die, we want people to accept it. We're in a risky business, and we hope that if anything happens to us, it will not delay the program. The conquest of space is worth the risk of life*". NASA took this testament seriously and proceeded with the program. From November 1967 to April 1968, three unmanned missions, Apollo 4, 5, and 6, were flown to test the hardware. Then, on October 11, 1968, Walter Schirra (commander), Donn Eisele (1930–1987), and Walter Cunningham (1932–2023) performed a first test of the lunar spacecraft while remaining in Earth orbit.

On the other side of the Iron Curtain, Mishin continued along the line drawn by Korolev. After the tragic end of Komarov, the *Soyuz*, the "*machine of the future*" according to his father, the *glavny konstruktor*, had been reworked from top to bottom and modified in the parts that had shown the greatest criticality, starting with the parachutes. The Soviets continued to prefer hard ground for re-entries to the soft waves of the sea because they had large prairies for landing, while they had difficulty finding a stretch of water that did not present almost insurmountable logistical problems and the risk of delivering the spacecraft into unwanted hands.

With the "close encounter" tests archived, it was time to experiment with actual docking. Taking advantage of the total automation philosophy imposed by Korolev, two unmanned *Soyuz* spacecraft were launched on October 30, 1967. They performed the docking brilliantly, but still showed numerous flaws, to the point that one of them had to be destroyed in flight. To disguise the true nature of the failed missions, they were renamed *Cosmos 186* and *188*. The Russian leopard kept once again its spots unchanged.

The exercise was repeated with *Cosmos 212* and *213*, which were launched on April 15 of the following year. This time everything went perfectly. The success made it possible to try again with a human crew. The problem to be solved was the format of the test, which involved the docking and transfer of one or two astronauts from one *Soyuz* to another. After extensive discussions, poisoned by partisan interests and ideological factors, including the selection of crew members, it was agreed to reduce the risks and send only one cosmonaut to meet a second unmanned spacecraft.

The forty-seven-year-old Georgy Beregovoy (1921–1995), Brezhnev's favorite because he was also Ukrainian, was chosen as commander of *Soyuz-3*. The launch took place in the afternoon of October 26, 1968. The day before, the empty *Soyuz-2* capsule had been launched, also from Baikonur. The rendezvous between the two spacecraft went well, but the docking failed due to trivial assembly errors of the guide lights and some inexperience of Beregovoy in controlling the vehicle. The landing four days later went perfectly. Overall, however, the mission was considered a failure.

The Apollo program, instead, proceeded quickly and smoothly toward the Moon. On December 21, 1968, Apollo 8 was launched with Frank Borman, James Lovell, and William Anders (1933–) on board, reaching lunar orbit and returning home safely. An epochal achievement that put America far ahead of its competitors.

It should be said that the mission had been planned with a different objective: to test the operation of the lander in Earth orbit. But since the LEM was not ready, NASA changed the program to boost the morale of a nation that had been wounded by a particularly difficult year.[9] In addition, James Webb had been informed by the CIA that the Soviets were organizing a lunar mission of their own. It was necessary to act in advance, perhaps forcing the adversary to make a misstep. "*We didn't feel sure that we could win it, but we felt sure we could compete*", he would recall then the courageous NASA administrator.

Webb still had the sarcastic words of a Missouri senator ringing in his ears, spoken during a Capitol Hill hearing after Leonov's spacewalk four years earlier. In his capacity as head of NASA, he had defended the organization by arguing that while the Russians were doing more spectacular things, the Americans had developed more sophisticated scientific equipment capable of performing, for example, superb photographic missions. The Senator, completely dissatisfied, had replied that these were just words scattered to the wind with propaganda skill, while "*the Russians were walking in space*". After years of hard work, the time had finally come to make the arrogant politician swallow his words. The Moon was now within reach of the Yankee astronauts.

---

[9] In addition to the Sixty-Eight protests, for Americans 1968 was marked by a major escalation of the Vietnam War, the assassination of Robert Kennedy, and the inability to support the Prague Spring.

Although they were now far behind, the Soviets continued on their way and managed to set another record with *Soyuz-4* and *5*, which were launched into orbit on January 14 and 15, 1969, respectively. The two spacecraft docked in orbit on January 16 and remained together for four and a half hours. Two cosmonauts, Aleksei Yeliseyev and Yevgeny Khrunov, transferred in EVA mode[10] to the *Soyuz-4*, which had been launched with only one passenger, Boris Volynov (1934–), while Vladimir Shatalov (1927–2021) remained alone in the *Soyuz-5*. Maneuver successful![11]

It was, in fact, the first step toward the realization of a space station assembled piece by piece in orbit and the demonstration of the feasibility of the in-flight docking necessary for the journey to the Moon. Von Braun was right. The Bolsheviks could not be defeated here. But they could be beaten for the conquest of the Moon. And indeed, on March 3, Pete Conrad, Richard Gordon and Alan Bean (1932–2018) aboard Apollo 9 successfully tested the Lunar Module left behind by Apollo 8. Forty-five days later, Thomas Stafford, John Watts Young and Eugene Cernan repeated the exercise in lunar orbit on Apollo 10. The moment of truth had arrived.

On July 16, 1969, at 1:30 p.m., Greenwich time, Neil Armstrong, Michael Collins, and Buzz Aldrin left Cape Kennedy for the Moon. Everything went according to plan: launch, parking in Earth orbit, jump to the Moon, entry into lunar orbit, and jettisoning of the Lunar Module. At 20 h and 18 min on July 20, the LEM Eagle landed on the lunar surface. "*Houston, this is Tranquility Base. The Eagle has landed*", reported Armstrong, who was in the Lunar Module together with Aldrin, while Collins remained on watch on the shuttle Columbia. "*Neil's voice was calm, confident, and most of all, clear*", Alan Shepard, the first American in space, would later say of the first American on the Moon. In Houston, there was a moment of absolute silence, then an explosion of uncontrollable joy. The American team had scored.

After six and a half hours of preparation, Armstrong exited the LEM and, with justifiable caution, descended the ladder. With bated breath, and the world watching with a delay of just over a second (the time it takes light to reach the Earth from the Moon), he stepped onto that "*magnificent desolation*" for the first time and uttered the famous phrase, surely prepared in advance: "*That's one small step for a man, one giant leap for mankind*", thus securing game, set and match.

---

[10] The Russians had not yet implemented in the *Soyuz* a system for transferring cosmonauts from one ship to another without the need for a spacewalk.

[11] The celebration of the astronauts' happy return was marred by the aforementioned attack on Brezhnev in which Leonov was involved. Almost a sign to the Soviets that good fortune had now betrayed them by turning to overseas competitors.

"*The rockets that have made spaceflight possible are an advance that, more than any other technological victory of the twentieth century, was grounded in science fiction*", wrote Isaac Asimov, the Russian-American science fiction writer. "*Finally, in 1969 science fiction writers and readers had the odd experience of watching astronauts land on the Moon and seeing something happen exactly as they had always imagined it would [...] One thing that no science fiction writer visualized, however, as far as I know, was that the landings on the Moon would be watched by people on Earth by way of television*". Live landings and then live wars to get the taxpayers involved and to be able to manipulate their minds.

The Moon would have to wait no longer. The Yankees had overwhelmingly won the race that fate had long mockingly waved in front of the Soviets' eyes. The giant *N1* rocket, tested in January and early July, had failed. It would never fly. The race was over. *Pravda* reported the American triumph with a small article on the front page, followed by three columns in the main section. The news could not be silenced, but neither should it be overemphasized. If anything, it was an opportunity to stress the USSR's great role in the space race:

Man's walk on the Moon will go down into the chronicles of the twentieth century as a marvelous event along with such interrelated wonderful achievements as the launching of the first man-made satellite, the first space flight by Yuri Gagarin, Aleksey Leonov's first walk in space, and the first launching of automatic spacecraft towards the Moon, Venus, and Mars.

It was necessary to protect the morale of the citizens of the immense country who had given so much of their resources, ingenuity, courage, and participation to win a contest whose stakes were, above all, national prestige, and which they had deluded themselves into believing they could win.

I think Russia had no chance to be ahead of the Americans under Sergei Korolev and his successor, Vasily Mishin. [...] Korolev was not a scientist, not a designer: he was a brilliant manager. Korolev's problem was his mentality. His intent was to somehow use the launcher he had [N1]. His philosophy was, let's not work by stages, but let's assemble everything and then try it. And at last it will work. There were several attempts and failures with Lunnik [unmanned Soviet Moon probes]. Sending man to the Moon is too complicated, too complex for such an approach. I think it was doomed from the very beginning.

Sergei Khrushchev's opinion, though authoritative, is flawed by his umbilical cord to Chelomei, one of Korolev's rivals, and somehow falsified by the observation that even after Apollo 11 and the cancelation of the lunar program due to lack of motivation, first the Soviets and then today's Russians remained a great power in space, rivals and leading partners for the rest of the world.

A pathetic note. James Webb, the man who led NASA's comeback, had resigned the year before the Moon landing, at the end of Johnson's presidency. A staunch and militant Democrat, he chose to retire after the election of Richard Nixon rather than face the inevitable spoils system with a new Republican president. Almost half a century later, America wanted to thank him by naming the second prestigious space telescope[12] after him, the JWST, which NASA launched on Christmas 2021.

There was no ecstatic and applauding crowd when Columbus knelt on the beach of what he thought was Cipango (Japan) and baptized that strip of land San Salvador (Holy Savior). Only a handful of scoundrels arrived on three secondhand caravels lent to an adventurer by the very Catholic Kings of Castile and Aragon. The enterprise that historians have adopted as a watchtower between the Renaissance and the Modern Age required a very modest investment. The conquest of space, instead, burned up rivers of money and ended in front of a worldwide audience, because it was not only an adventure, but also an investment and a tug-of-war between two giants, one of which had with great ideals and feet of clay.

The race to the Moon had cost a lot, but it had paid off, especially in image and new technology. Now that this goal had been achieved, space demanded a new and greater effort to fulfill man's destiny. It was necessary to pool forces and put an end to the race for the most powerful rocket. The now violated Moon helped to carry out a controlled disarmament and to join the efforts for the colonization of space.

And it really happened according to the words of Charles Bolden, NASA administrator, and John Holdren, senior advisor to President Barack Obama, in a document posted in 2014: "*With a partnership that includes 15 nations and with 68 nations currently using the ISS [International Space Station] in one way or another, this unique orbiting laboratory is a clear demonstration of the benefits to humankind that can be achieved through peaceful global cooperation*". A laboratory that, until recently, could only be reached by *Soyuz*.

Then came Trump's statements with which we opened these pages. But that's another story, quite different, it seems, from the story of the paladins and the Moors at the lunar Roncesvalles. A story prophetically sketched by Verne in the opening of his *From the Earth to the Moon*:

---

[12] The first, the HST, is named after Edwin Powell Hubble, the astronomer who discovered the nature of galaxies and the law of expansion of the universe.

During the Federal war in the United States a new and very influential club was established in the city of Baltimore, Maryland. It is well known with what energy the military instinct was developed amongst that nation of shipowners, shopkeepers, and mechanics. Mere tradesmen jumped their counters to become extempore captains, colonels, and generals without having passed the Military School at West Point; they soon rivalled their colleagues of the old continent, and, like them, gained victories by dint of lavishing bullets, millions, and men. But where Americans singularly surpassed Europeans was in the science of ballistics, or of throwing massive weapons by the use of an engine; not that their arms attained a higher degree of perfection, but they were of unusual dimensions, and consequently of hitherto unknown ranges. The English, French, and Prussians have nothing to learn about flank, running, enfilading, or point blank firing; but their cannon, howitzers, and mortars are mere pocket pistols compared with the formidable engines of American artillery. This fact ought to astonish no one. The Yankees, the first mechanicians in the world, are born engineers, just as Italians are musicians and Germans meta-physicians.

Is there no hope? "*Homo homini lupus*" (Man is wolf to man) forever? "*It was a wonderful night, such a night as is only possible when we are young, dear reader. The sky was so starry, so bright that, looking at it, one could not help asking oneself whether ill-humored and capricious people could live under such a sky*", asked Fyodor Dostoyevsky at the beginning of *White Nights*. A rhetorical question. We all know it's possible! History teaches us this. But, in addition to evil men, there are those, "*li omini boni*" (good men) invoked by Leonardo da Vinci, who like to learn, look at the stars, and climb to the sky: Dr. Jekyll as well as Mr. Hyde. A pinch of hope in the pessimism of reason.

Printed in the United States
by Baker & Taylor Publisher Services